C000268962

Collins

Weather

ALMANAC

A GUIDE TO
2024

Storm Dunlop

Published by Collins
An imprint of HarperCollins Publishers
Westerhill Road
Bishopbriggs
Glasgow G64 2QT
www.harpercollins.co.uk

HarperCollins Publishers
Macken House
39/40 Mayor Street Upper, Dublin 1
D01 C9W8 Ireland

A catalogue record for this book is available from the British Library

ISBN 978-0-00-861795-0
10 9 8 7 6 5 4 3 2 1

Printed in the UK using 100% Renewable Electricity at
CPI Group (UK) Ltd

If you would like to comment on any aspect of this book,
please contact us at the above address or online.
e-mail: collins.reference@harpercollins.co.uk

 facebook.com/CollinsAstronomy
 @CollinsAstro

MIX
Paper from
responsible sources
FSC™ C007454

This book is produced from independently certified
FSC™ paper to ensure responsible forest management.

For more information visit: www.harpercollins.co.uk/green

Contents

Introduction

Our variable weather

Anyone living in the British Isles hardly needs to be told that the weather is extremely variable. Despite what politicians and the media may say, the extreme weather sometimes experienced in Britain is not 'unprecedented'. Britain has always experienced extreme windstorms, snowfall, rainfall, thunderstorms, flooding and such major events. And always will. Such events may be unusual, and 'not within living memory of the oldest inhabitant', but, overall, the weather always exhibits such extremes. We may have learned from events such as the East Coast floods of 1953, and constructed the Thames Barrier and the barrier on the River Hull, but a North Sea surge will occur again. The Somerset Levels have been flooded many times in the past, so the flooding in 2012 and 2014 was not that extraordinary. They were flooded in 1607, so have a long history of flooding. Global warming may produce more occasions when extreme events occur, but no meteorologists can predict when these may happen.

The weather in Britain – its climate – is basically a maritime climate, determined by the proximity of these islands to the Atlantic Ocean. It is largely determined by the changes resulting from incursions of dry continental air from the Eurasian landmass to the east, contrasting with the prevailing moist maritime air from the Atlantic Ocean to the west. The general mildness of the climate, in comparison with other locations at a similar latitude, has often been ascribed to 'the Gulf Stream'. In fact, the Gulf Stream exists only on the western side of the Atlantic, along the East Coast of America, and the warm current off the coast of Britain is correctly known as the North Atlantic Drift. In reality, it is not solely the warmth of the oceanic waters that creates the mild climate.

The Rocky Mountains in North America impede the westerly flow of air and create a series of north/south waves that propagate eastwards (and actually right round the world). These waves cause north/south oscillations of the jet stream (which, in our case, is the 'Polar Front' jet stream) that, in turn, controls the progression of the low-pressure areas (the depressions) that travel from west to east and create most of the changeable weather over Britain. The jet stream 'steers' the depressions, sometimes to the north of the British Isles and sometimes far to the south. It may also assume a strong flow in longitude (known as a 'meridional' flow) leading to a blocking situation, where depressions cannot move eastwards and may come to a halt or be forced to travel far to the north or south. Such a block (especially in winter) may draw frigid air directly from the Arctic Ocean or continental air from the east (such air is usually described as 'from Siberia', because it often originates from that region). More rarely, such blocks draw warmer air from the Mediterranean over the country.

The location of the jet stream itself is governed by something known to meteorologists as the North Atlantic Oscillation (NAO). Basically, this may be thought of as the distribution of pressure between the Azores High over the central Atlantic and the semi-permanent Icelandic Low. These are semi-permanent features of the flow of air around the globe, and are known to meteorologists as 'centres of action'. The average route of the jet stream is to the north of the British Isles, so the whole country is subject to the strong westerlies that prevail at these latitudes. Northern regions of the British Isles tend to be mainly subject to the depressions arriving from the Atlantic and the weather that accompanies them. So overall, the climate of the British Isles may be described as: warmer and drier in the south and east, wetter in the west and north.

When the NAO has a 'positive' index, with high pressure and warm air in the south compared with low pressure and low temperatures in the north, depressions are steered north of the British Isles. Although most of the country experiences windy and wet weather, the west and north tends to be affected most, with southern and eastern England warmer and drier. When the NAO has a 'negative' index, the Azores High is displaced towards the north and the jet stream tends to show strong meanders towards the south. Frequently there is a slowly moving low pressure area over the near (north-western) Continent or over the north-eastern Atlantic. The jet stream wraps around this, before turning back to the north. Blocks are, however, most frequent in the spring.

Although the weather across Britain is always changeable, there are certain characteristic areas of the country, and the weather associated with them. Details of these distinct areas, and the weather that you are likely to experience within them, are given on pages 213–232.

Climatic regions of the British Isles
It is appropriate to deal with the climates of the British Isles by discussing eight separate regions, which are:

1 South-West England and the Channel Islands (page 214)
2 South-East England and East Anglia (page 216)
3 The Midlands (page 218)
4 North-West England and the Isle of Man (page 220)
5 North-East England and Yorkshire (page 222)
6 Wales (page 224)
7 Ireland (page 226)
8 Scotland (page 229)

Sunrise/Sunset & Moonrise/Moonset
For four dates within each month, tables show the times at which the Sun and Moon rise and set at the four capital cities in the United Kingdom: Belfast, Cardiff, Edinburgh and London. For Edinburgh and London, the locations of specific observatories are used, but not for Belfast and Cardiff, where more general locations are employed. Although the exact time of Sunrise/Sunset and Moonrise/Moonset at any observer's location depends on their exact position, including their latitude and longitude and height above sea level, the times shown will give a useful indication of the timing of the events. However, calculation of rising and setting times is complicated and strongly depends on the location of the observer. Note that all the times in this book are calculated astronomically, using what is known as Universal Time (UT), sometimes known as Coordinated Universal Time (UTC). This is identical to Greenwich Mean Time (GMT). The times given do not take account of British Summer Time (BST).

Note that on rare occasions, no time is shown for Moonrise. This is because the Moon rose **the day before**. The time of Moonrise is therefore shown as '–' and a second line shows the time and azimuth at which the moon rose the day before. A similar problem sometimes occurs with the time of Moonset, which actually occurs **the following day.** The appropriate time and azimuth are thus shown in a similar fashion.

Some effects are quite large. For example, at the summer solstice in 2024, sunrise is some 38 minutes later at Edinburgh than it is at Lerwick in the Shetlands. Sunset is some 21 minutes earlier on the same date.

Edinburgh & Lerwick: Comparison of sunrise & sunset times
A comparison of the timings of sunrise and sunset at Lerwick (latitude 60.16 N) and Edinburgh (latitude 55.9 N) in the following table gives an idea of how the times change with latitude. The timings are those at the equinoxes (March 20 and September 22 in 2024) and the solstices (June 20 and December 21 in 2024). It is notable, of course, that, although the times are different, the azimuth of sunrise is identical at both equinoxes, because the Sun is then crossing the celestial equator. At the solstices, both times and azimuths are different between the two locations. Azimuth is measured in degrees from north, through east, south and west, and then back to north.

Location	Date	Rise	Azimuth	Set	Azimuth
Edinburgh	20 Mar 2024 (Wed)	06:14	89	18:27	272
Lerwick	20 Mar 2024 (Wed)	06:05	88	18:20	272
Edinburgh	20 Jun 2024 (Thu)	03:26	43	21:02	317
Lerwick	20 Jun 2024 (Thu)	02:39	34	21:34	326
Edinburgh	22 Sep 2024 (Sun)	05:59	89	18:11	271
Lerwick	22 Sep 2024 (Sun)	05:50	88	18:03	271
Edinburgh	21 Dec 2024 (Sat)	08:42	134	15:40	226
Lerwick	21 Dec 2024 (Sat)	09:08	141	14:58	219

Apart from the times, this table (and the monthly tables) also shows the azimuth of each event, which indicates where the body concerned rises or sets. These azimuths are given in degrees, and the table given here shows the azimuths for various compass points in the eastern and western sectors of the horizon.

Table of azimuths

Degrees	Compass point		
Eastern horizon		*Western horizon*	
0°	N	202° 30'	SSW
45°	NE	225°	SW
67° 30'	ENE	247° 30'	WSW
90°	E	270°	W
112° 30'	ESE	292° 30'	WNW
135°	SE	315°	NW
157° 30'	SSE	337° 30'	NNW

The actual latitude and longitude used in the calculations are shown in the following table. It will be seen that the altitudes of the observatories in Edinburgh (Royal Observatory Edinburgh, ROE) and London (Mill Hill Observatory) are quite considerable (that for ROE is particularly large) and these altitudes will affect the rising and setting times, which are calculated to apply to observers closer to sea level. (Generally, close to the observatories, such rising times will be slightly later, and setting times slightly earlier than those shown.)

Latitude and longitude of UK capital cities

City	Longitude	Latitude	Altitude
Belfast	5°56'00.0" W	54°36'00.0" N	3 m
Cardiff	3°11'00.0" W	51°30'00.0" N	3 m
Edinburgh (ROE)	3°11'00.0" W	55°55'30.0" N	146 m
London (Mill Hill)	0°14'24.0" W	51°36'48.0" N	81 m

Twilight

For each individual month, we give details of sunrise and
sunset times (and Moonrise and Moonset times), together with
the azimuths (which give an idea of where the rising or setting
takes place – see previous page) for the four capital cities of
the regions of the United Kingdom. But twilight also varies
considerably from place to place, so the monthly diagrams
here show the duration of twilight at those four cities. During
the summer, twilight may persist throughout the night. This
applies everywhere in the United Kingdom, so two additional
yearly twilight diagrams are included (on pages 259 and 261):
one for Lerwick in the Shetlands and one for St Mary's in the
Scilly Isles. Although the hours of complete darkness increase
as one moves towards the equator, it will be seen that there is
full darkness nowhere in the British Isles at midsummer.

There are three recognised stages of twilight: *civil twilight*,
when the centre of the Sun is less than 6° below the horizon;
nautical twilight, when the Sun is between 6° and 12° below the
horizon; and *astronomical twilight*, when the Sun is between 12°
and 18° below the horizon. Full darkness occurs only when the
Sun is more than 18° below the horizon. The time at which civil
twilight begins is sometimes known in the UK as 'lighting-up
time'. Stars are generally invisible during civil twilight. During
nautical twilight, the very brightest stars only are visible. (These
are the stars that were used for navigation, hence the name
for this stage.) During astronomical twilight, the faintest stars
visible to the naked eye may be seen directly overhead, but are
lost at lower altitudes. They become visible only once it is fully
dark. The diagrams show the duration of twilight at the various

cities. Of the locations shown, during the summer months there is astronomical twilight at most of the locations, except at Lerwick, but there is never full darkness during the summer anywhere in the British Isles.

The diagrams also show the times of New and Full Moon (black and white symbols, respectively). As may be seen, at most locations during the year, roughly half of New and Full Moon phases may come during daylight. For this reason, the exact phase may be invisible at one location, but be clearly seen elsewhere in the world. The exact times of the events are given in the diagrams for each individual month.

Twilight diagrams for each of the four capital cities are shown every month, and full yearly diagrams are shown on pages 259–261.

Also shown each month is the phase of the Moon for every day, together with the age of the Moon, which is counted from New Moon.

Weather extremes are shown for each month, and the locations plotted on the appropriate maps.

Where possible, a notable weather event that occurred in the month concerned is described in detail. For some months, this is replaced by a weather topic of more general interest.

The seasons

By convention, the year has always been divided into four seasons: spring, summer, autumn and winter. In the late eighteenth century, an early German meteorological society, the Societas Meteorologica Palatina, active in the Rhineland, defined the seasons as each consisting of three whole months, beginning before the equinoxes and solstices. So spring consisted of the months of March, April and May; summer of June, July and August; autumn of September, October and November, and winter of December, January and February. There has been a tendency by meteorologists to follow this convention to this day, with winter regarded as the three calendar months with the lowest temperatures in the northern hemisphere (December, January and February) and summer those with the warmest (June, July and August). Astronomers, however, regard the seasons as lasting three months, but centered on the dates of the equinoxes and solstices (20 March, 22 September and 20 June, 21 December in 2024).

Some ecologists tend to regard the year as divided into six seasons. Analysis of the prevailing weather types in Britain, however, suggests that there are five distinct seasons. (The characteristic weather of each is described in the month in which the season begins.) Although, obviously, the seasons cannot be specified as starting and ending on specific calendar dates, it is useful to identify them in this way. So in Britain, we have:

Early winter
20 November to 19 January (see page 181)

Late winter and early spring
20 January to 31 March (page 17)

Spring and early summer
1 April to 17 June (page 65)

High summer
18 June to 9 September (page 99)

Autumn
10 September to 19 November (page 148)

The general situation and the weather in the year 2022 are described (with specific reference to the various regions) on the next two pages. This is the latest year for which reliable information is available. The year 2023 is still in progress as these words are written, so only extreme events in early 2023 are mentioned.

The Weather in 2022

Five named storms affected Britain in 2022. These were all (as expected) in the winter period. Two (Storms Malik and Corrie) occurred in January, causing unsettled weather, and no fewer than three (Storms Dudley, Eunice and Franklin) affected the country in a single week (16 to 21 February). The Met Office issued a red warning of high winds for southern Wales and the south-west for Storm Eunice, which was extremely destructive. A gust of 106 knots (122 mph/196 kph) was recorded at Needles Old Battery on the Isle of Wight during this storm, setting a new gust record for Britain.

January was generally dry, but February (apart from suffering from the storms, especially Storm Eunice) was wet and mild, with few frosts, especially in the south of England.

Spring tended to be warmer than average, although the beginning of March continued the unsettled weather of the end of February. Clear skies produced warmer than usual daytime temperatures, but some cold nights and late frosts, particularly in central and northern England. Rainfall was generally lower than normal, especially in western Scotland, and the north-west of England and Wales. Overall, it was sunnier than normal, particularly in Scotland and Northern Ireland.

Summer was much warmer and drier than average, especially in eastern counties of England. Many locations experienced extremely high temperatures, with Coningsby in Lincolnshire setting a UK record of 40.3°C on 19 July. Six sites experienced temperatures over 40°C and no fewer than 28 locations exceeded the previous record temperature of 38.7°C. It was the hottest summer ever for the UK as a whole, and is generally thought to be caused by climate change and global warming.

The autumn of 2022 was slightly warmer than average. (It was the third warmest autumn on record for the UK.) Although September began fine and warm, the weather then became unsettled, becoming cooler. However, both October and November were warmer than average. November was particularly wet in the south, with some locations having twice their usual amount of rain. The far north of Scotland and Northern Ireland were much drier.

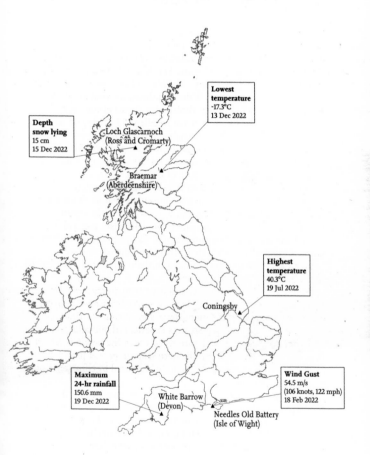

Depth snow lying
15 cm
15 Dec 2022

Loch Glascarnoch
(Ross and Cromarty)

Lowest temperature
-17.3°C
13 Dec 2022

Braemar
(Aberdeenshire)

Highest temperature
40.3°C
19 Jul 2022

Coningsby

Maximum 24-hr rainfall
150.6 mm
19 Dec 2022

White Barrow
(Devon)

Needles Old Battery
(Isle of Wight)

Wind Gust
54.5 m/s
(106 knots, 122 mph)
18 Feb 2022

January

Introduction

January sees some of the coldest temperatures of the year. Braemar, the village in the Scottish Highlands, about 93 km west of Aberdeen, has twice recorded the lowest temperature in the British Isles (-27.2°C), first on 11 February 1895 and since then on 10 January 1982. Only Altnaharra in Sutherland has ever recorded a similar temperature (on 30 December 1995). Generally, Scotland sees extensive snow cover in January, particularly important at the ski centres at Cairngorm Mountain, Glencoe Mountain, Glenshee, Nevis Range and The Lecht, although global warming threatens to decrease the coverage of snow. Deep snow has become less frequent and the centres have had to invest in snow-making cannons.

The weather in January is often dominated by cold easterly winds. These arise because of the cold anticyclone over the near continent that builds up during most winters. This is often an extension of the great wintertime Siberian High that dominates the weather over Asia and creates a cold, dry airflow over the eastern side of Asia. Circulation round this anticyclone tends to bring a cold airflow across the North Sea. In crossing the sea, the air gains moisture and this often results in snowfall along the eastern coast of Britain. At times, a blocking situation may arise with a major high-pressure area over Scandinavia. This brings extremely cold Arctic air down over the British Isles and, depending on how long the block persists, may give rise to a persistent spell of low temperatures.

Snowfall is, however, very frequent farther south, and the exceptional snowfall in 1947, which was undoubtedly the winter that saw the greatest fall of snow over Britain, did not begin until quite late in the month of January. Although there had been some snow earlier, in December and early January, it had melted by the middle of the month, and there were unseasonably high temperatures across the country. The temperature then dropped and there were frosts at night from 20 January (the nominal beginning of the late winter, early spring season). Snow-bearing clouds began to move into the south-west of England on 22 January. There was heavy snowfall with blizzard conditions in the West Country. Even the Scilly Isles saw a slight covering of snow, amounting to a few centimetres in depth – an almost unprecedented event. The following days saw heavy snowfall extend right across all of England and Wales before spreading into Scotland. There were seemingly relentless snowstorms over the next few weeks, which left England and Wales, up as far as the Scottish Borders, buried beneath a deep blanket of snow, and movement by rail and road almost completely paralysed. Somewhat ironically, although snow fell somewhere in the United Kingdom on every day from 22 January until the middle of March, Scotland escaped the worst storms. After some three months of northerly and easterly winds, by 10 March, warm southerly and south-westerly winds started to affect the West Country. Not only did the warm airstream bring dense fogs, it also produced heavy rainfall, which in turn led to floods as the rain ran off the frozen ground. The situation was worsened by the gradual thawing of the immense snowpack, leading to even more extreme flooding. By 13 March, even the rivers in East Anglia were about to burst their banks.

Only the winter of 1962–63 saw a longer period during which snow persisted, and much lower temperatures, but the amount of snow that fell in 1947 was the most extreme ever recorded for the United Kingdom.

Late winter and early spring season – 20 January to 31 March
This season tends to exhibit long spells of settled conditions. These may be of very cold weather, characterised by Arctic air, introduced by a northerly airflow, and thus forming the main period of winter. Although in some years the weather may take on the character of an extended spring, such conditions are less frequent than those with low temperatures. There may be long spells of wet, westerly conditions, but these tend to be less common than spells with cold northerly or easterly winds. However, the wet westerlies suddenly become less frequent after about 9 March, and indeed westerly weather then becomes very uncommon.

Weather Extremes in January

Country	Temp.	Location	Date
Maximum temperature			
England	17.6°C	Eynsford (Kent)	27 Jan 2003
Wales	18.3°C	Aber (Gwynedd)	10 Jan 1971 27 Jan 1958
Scotland	18.3°C	Aboyne (Aberdeenshire) Inchmarlo (Kincardineshire)	26 Jan 2003
Northern Ireland	16.4°C	Knocharevan (Co. Fermanagh)	26 Jan 2003
Minimum temperature			
England	-26.1°C	Newport (Shropshire)	10 Jan 1982
Wales	-23.3°C	Rhayader (Powys)	21 Jan 1940
Scotland	-27.2°C	Braemar (Aberdeenshire)	10 Jan 1982
Northern Ireland	-17.5°C	Magherally (Co. Down)	1 Jan 1979

Country	Pressure	Location	Date
Maximum pressure			
Scotland	1053.6 hPa	Aberdeen Observatory	31 Jan 1902
Minimum pressure			
Scotland	925.6 hPa	Ochtertyre (Perthshire)	26 Jan 1884

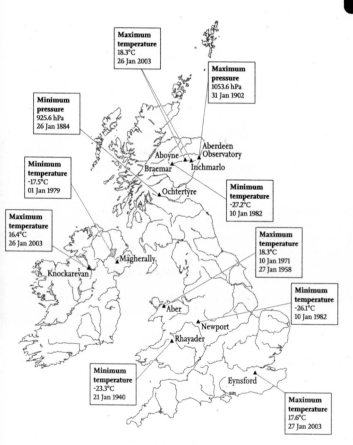

Maximum temperature
18.3°C
26 Jan 2003

Maximum pressure
1053.6 hPa
31 Jan 1902

Minimum pressure
925.6 hPa
26 Jan 1884

Minimum temperature
-17.5°C
01 Jan 1979

Maximum temperature
16.4°C
26 Jan 2003

Aberdeen Observatory
Aboyne
Braemar
Inchmarlo
Ochtertyre

Minimum temperature
-27.2°C
10 Jan 1982

Maximum temperature
18.3°C
10 Jan 1971
27 Jan 1958

Magheraly
Knockarevan

Minimum temperature
-26.1°C
10 Jan 1982

Aber
Newport
Rhayader

Minimum temperature
-23.3°C
21 Jan 1940

Eynsford

Maximum temperature
17.6°C
27 Jan 2003

The Weather in January 2023

Observation	Location	Date
Max. temperature 15.8°C	Dyce (Aberdeenshire)	24 January
Min. temperature -10.4°C	Drumnadrochit (Inverness-shire)	19 January
Most rainfall 100.2 mm	Maerdy Water Works (Mid Glamorgan)	11 January
Greatest snow depth 34 cm	Loch Glascarnoch (Ross & Cromarty)	18 and 19 January

January was a month of two halves. The beginning of the month continued with the mild, unsettled weather that had ended December 2022. This was caused by a series of deep depressions interrupted by ridges of high pressure over the Atlantic, while the jet stream was weaker than usual and located towards Europe. The highest minimum temperature occurred on the 5th, with St Marys Airport on the Isles of Scilly recording a minimum of 11.7°C.

With this came heavy rain, especially from the 10th to 13th. Some places received their average monthly rainfall for January within the first two weeks. Impacts on travel included road closures and flooded rail lines in South Wales, and the midlands and Somerset Levels also experienced flooding. Maerdy Water Works in Mid-Glamorgan received over 100 mm of rainfall on the 11th. Wind speeds were also high during this period, with northern Scotland seeing gusts over 50 mph (80.5 kph/43.4 knots) across wide areas.

The second half of the month saw temperatures fall and rainfall decrease sharply as winds turned northerly. Parts of Scotland recorded significant snowfall (34 cm fell at Loch Glascarnoch on the 18th and 19th), as did Cornwall and North Wales when heavy snow showers arrived from the Irish Sea.

Once the worst of the cold snap passed, the weather for the end of the month was more benign. By the 23rd, high pressure had established itself over the south of the UK, with colder weather persisting and several fog warnings issued. High volumes of surface water remained from earlier in the month in some places. In the north of England and Scotland the highest maxima of the month were witnessed (15.8°C at Dyce, Aberdeenshire on the 24th), and the milder weather saw some rain return.

Overall, despite these swinging extremes, mean temperatures were close to average. The overall mean was 0.4°C above the 1991–2020 average, with northern parts of England and Wales the most so, but nowhere saw anomalies of greater than 1°C. Rainfall was also close to average, but this doesn't tell the whole story; parts of western England and Wales received 150 per cent of their normal monthly rainfall, while eastern Scotland and north-east England received only 50 per cent. England saw its second sunniest January on record, and the UK as a whole saw 133 per cent of its average sun for the month.

Sunrise and Sunset 2024

Location	Date	Rise	Azimuth °	Set	Azimuth °
Belfast					
	1 Jan (Mon)	08:46	131	16:08	229
	11 Jan (Thu)	08:41	128	16:22	232
	21 Jan (Sun)	08:31	125	16:39	235
	31 Jan (Wed)	08:16	120	16:59	240
Cardiff					
	1 Jan (Mon)	08:18	128	16:14	232
	11 Jan (Thu)	08:15	125	16:26	235
	21 Jan (Sun)	08:06	122	16:42	238
	31 Jan (Wed)	07:53	118	17:00	243
Edinburgh					
	1 Jan (Mon)	08:44	133	15:49	227
	11 Jan (Thu)	08:38	130	16:03	230
	21 Jan (Sun)	08:26	126	16:22	234
	31 Jan (Wed)	08:10	121	16:43	239
London					
	1 Jan (Mon)	08:07	128	16:02	232
	11 Jan (Thu)	08:04	126	16:14	235
	21 Jan (Sun)	07:55	122	16:30	238
	31 Jan (Wed)	07:42	118	16:47	242

Note that all times are in Universal Time (UT), otherwise known as Greenwich Mean Time (GMT). These times do not take Summer Time (BST) into account.

Moonrise and Moonset 2024

Location	Date	Rise	Azimuth °	Set	Azimuth °
Belfast					
	1 Jan (Mon)	22:12	77	11:36	288
	11 Jan (Thu)	09:34	142	15:46	220
	21 Jan (Sun)	12:08	41	05:35	318
	31 Jan (Wed)	23:38	104	10:04	262
Cardiff					
	1 Jan (Mon)	22:06	78	11:19	286
	11 Jan (Thu)	08:57	137	15:58	224
	21 Jan (Sun)	12:20	45	05:01	313
	31 Jan (Wed)	23:21	103	09:56	262
Edinburgh					
	1 Jan (Mon)	21:58	76	11:28	288
	11 Jan (Thu)	09:35	145	15:21	217
	21 Jan (Sun)	11:45	38	05:34	320
	31 Jan (Wed)	23:29	104	09:52	262
London					
	1 Jan (Mon)	21:53	77	11:07	287
	11 Jan (Thu)	08:46	137	15:45	224
	21 Jan (Sun)	12:07	45	04:49	313
	31 Jan (Wed)	23:09	103	09:44	263

Note that all times are in Universal Time (UT), otherwise known as Greenwich Mean Time (GMT). These times do not take Summer Time (BST) into account.

Twilight Diagrams 2024

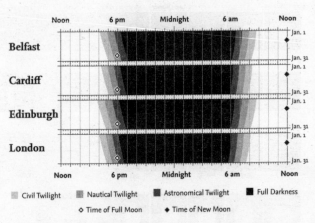

The exact times of the Moon's major phases are shown on the diagrams opposite.

Noctilucent clouds
Clouds seen in midsummer in the middle of the night (the name means 'night-shining') and (for Britain) in the general direction of the North Pole. These clouds (NLC) are seen when the observer is in darkness, but the clouds – which are the highest in the atmosphere at about 185 kilometres, far above all other clouds – are still illuminated by the Sun, which is below the horizon. They consist of ice crystals, believed to form around meteoritic dust arriving from space. They also occur in the southern hemisphere, although because of the distribution of land masses, are commonly seen only by observers in the Antarctic Peninsula (see pages 112 and 113).

The Moon's Phases and Ages 2024

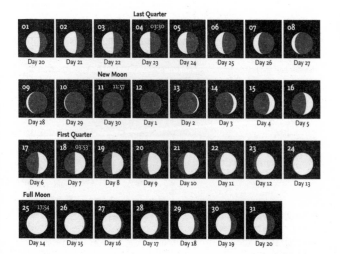

Aurora

A luminous event occurring in the upper atmosphere between approximately 100 and 1000 km. It arises when energetic particles from the Sun raise atoms to higher energy levels. When the atoms drop back to their original energy level, they emit the characteristic green and red shades that are visible to human eyes (from oxygen and nitrogen, respectively).

January – In this month

1 January 1922 – Weather Shipping, later to be known as the Shipping Forecast, was broadcast for the first time by the Met Office. A telegraph warning service had been established in 1861 by Vice-Admiral Robert FitzRoy, and localised twice-daily wireless transmission began in 1921. In 1925, transmission was taken up by the BBC.

❄

1 January 2022 – the warmest New Year's Day on record was reported. Temperatures of 16.2°C were recorded in St James's Park in central London.

❄

6 January 1928 – the River Thames burst its banks in London. The following day the moat at the Tower of London, which had been drained in 1843 and planted with grass, was completely refilled by the river.

❄

11 January 1954 – George Cowling became the first weather presenter to deliver the forecast on television, standing in front of a map. He made his debut as the BBC aimed to change the way the weather was presented, with more focus on the continuity of the forecast day to day.

J

12 **January 1899** – The schooner Forest Hall foundered off the coast of Devon, and thirteen crew members and five apprentices were rescued by the Lynmouth Lifeboat.

25 **January 1990** – the Burns Day Storm made landfall in Ireland before continuing east to Scotland, Denmark and Belgium. Winds of up to 75 mph were recorded, and the storm damage cost UK insurers £3.37 billion.

31 **January 1918** – A misty Scottish night contributed to a series of accidental collisions between two Royal Navy submarines and five warships. Both submarines were lost, with over a hundred lives. The other five ships sustained varying degrees of damage.

31 **January 1953** – A combination of a high spring tide and a severe windstorm caused the North Sea flood, a storm surge that hit the Netherlands, Belgium, England and Scotland. Most sea defences facing the surge were overwhelmed and, in the UK alone, 160,000 acres of land up to 5.6 m above mean sea level was flooded. Hundreds of people perished.

The Night of the Big Wind

6 January 1839 is known in Ireland as 'Oiche na Gaiothe Moire' – the 'Night of the Big Wind'. The weather preceding the windstorm was remarkably quiet. There had been a fall of snow, but the wind dropped and it was relatively calm.

Ireland bore the brunt of one of the deepest depressions to have approached these islands from the Atlantic. (Accurate measurements of the central pressure were not available at that time.) Naturally, the west coast first encountered the fury of the storm and vast amounts of seawater were cast ashore. Fish were later found many miles inland. By the middle of the night, winds, estimated to be up to 240 kph, were roaring across the relatively flat midland belt of Ireland, unimpeded by friction with the surface and any windbreaks. Because the storm hit during the night, relatively few people were killed, although the death

The loss of the Pennsylvania New York packet ship; the Lockwoods emigrant ship; the Saint Andrew packet ship; and the Victoria from Charleston, near Liverpool, during the hurricane on Monday and Tuesday 7 and 8 January 1839.

toll amounted to about 300. Many large structures were swept away, including windmills, trees, ancient monuments, barns and factories. Tombstones were even uprooted from cemeteries. Some were even found as much as a mile away from their origin. Many thousands were left homeless, and many found that their life savings, hidden in thatched roofs had been lost, and were thus destitute. Few birds survived, and the following spring is notable for the absence of birdsong.

The 'Night of the Big Wind' was long remembered in Ireland, and indeed when the official state pension system was introduced in 1909, the oldest residents, unable to prove their age at a time without documentation, were asked if they remembered the 'Night of the Big Wind'. If so, they were obviously over 70 years old.

DREADFUL HURICANE.—ALARMING FIRE.—One of the most appalling and destructive storms that ever visited our city occurred last night. It commenced about ten o'clock, and continued to blow with tremendous violence until about six o'clock this morning, when its fury somewhat abated. The damage done in the unroofing of houses, falling of chimneys, uprooting of trees, &c., must have been immense. One instance may give some idea of the fierceness of the storm:—A young man, in turning the corner of Sackville-street, in Britain-street, was blown off his feet, dashed against a lamp-post, and had his leg fractured in a shocking manner. A large tree was blown down at the corner of Rutland-square, near North Frederick-street. The horrors of the tempest are augmented by the additional calamity of fire. We have mentioned elsewhere, that a fire occurred in the Bethesda, on Saturday night. The same unfortunate building was last night the scene of a dreadful conflagration, which soon extended to the adjoining buildings. It is impossible to give a description of the awful appearance of the fire, in the middle of the hurricane. The streets were nearly impassable from the immense quantity of sparks which were driven about by the wind. The Bethesda and the school adjoining are complete wrecks; and the storm being too violent to allow the engines to be worked with effect, the fire continues to increase, and the houses next to them, in Dorset-street, are now igniting. It is said, that, if the storm does not soon abate, recourse will be had to artillery, to blow down the houses in danger, and stop the devouring progress of the conflagration.—*Register, of Monday.*

A newspaper report of the Night of the Big Wind. This description illustrates the scale of damage caused.

February

Introduction

Although in many rural areas (in southern England in particular), February has earned the nickname 'February Filldyke', in reality February may be extremely dry. February is, in fact, the one calendar month that is most likely to experience no rainfall whatsoever. As such, it sometimes shows spring-like conditions, which warrants its inclusion in the 'late winter, early spring' season (page 11). It has been found that a warm February in Scotland is a sign that the mean annual temperature in Scotland will probably be warm. However, a warm February may not be welcome for agriculture. Warmth indicates that precipitation will be in the form of rain, rather than snow. Rain in February is generally detrimental to seed and grass, but snow protects the ground and keeps it warm.

Yet spells of very cold weather may occur, especially if, as is often the case in January, a blocking situation occurs, especially a blocking anticyclone over Scandinavia, bringing frigid air down from the Arctic or similarly cold air from the semi-permanent cold anticyclone that builds up over Siberia in winter. The latter was the case with the exceptionally cold spell in 2018, nicknamed 'The Beast from the East' by the media, which began late in the month on 22 February, and continued until at least the end of the month. This particular cold wave originated in an exceptional anticyclone, named Anticyclone Hartmut, which transported frigid Siberian air west over Europe. In crossing the North Sea, the air gained a large amount of humidity, which produced extremely heavy snowfall that spread west over almost the whole of the United Kingdom and Ireland.

Somewhat similar conditions occurred in the middle of March (17 and 18 March), although this second cold spell did not last particularly long, and was not as severe. (The media sometimes called it the 'Mini Beast from the East'.)

The weather in February often shows a division between the north and south of the country, with cold temperatures and relatively dry conditions prevailing in the north of England and over Scotland, but much milder, wetter weather in the south, sometimes with exceptional rainfalls – hence the 'February

Filldyke' nickname. The 'Beast from the East' just mentioned not only brought cold conditions to Britain, but the frigid air it brought from the east then had a great effect on the humid air introduced from the west by Storm Emma. (This storm was actually named by the meteorological services of France, Spain and Portugal, rather than the name being taken from the list prepared by the Met Office and Met Eireann.)

Storm Emma originated in the Azores and was a typical deep depression, transporting a lot of warm moist air and accompanied by high winds. When the storm encountered the frigid air introduced over the British Isles by Anticyclone Hartmut, the warm, moist air was forced up over the cold air at the surface and Emma's moisture was deposited as snowfall. The snow depth reached as much as 57 cm in places, although a depth of 50 cm was widespread across the country. South-west England and south Wales were worst affected. Storm Emma also brought a renewed incursion of Arctic air and resulting low temperatures over much of the United Kingdom.

The effects of this collision between Hartmut and Emma were not confined to the British Isles. Snow fell along the French Riviera and in Italy. Even Barcelona in Spain saw snowfall – an unheard-of event for the region. The collision of the two systems also produced some exceptional winds, most notably a gust of 228 kph (142 mph) at Mont Aigoual in southern France on 1 March 2018.

Anemometer

Any device that measures wind speeds, generally in a horizontal direction. There are various types. The form most commonly seen is probably the type that has three rotating cups. Other versions use propellors, differences in pressure or the transmission of sound or heat. Certain devices (particularly sonic anemometers) are able to measure the vertical motion of the air, as well as motion in a horizontal direction.

Weather Extremes in February

Country	Temp.	Location	Date
Maximum temperature			
England	21.2°C	Kew Gardens (London)	26 Feb 2019
Wales	20.8°C	Porthmadog (Gwynedd)	26 Feb 2019
Scotland	18.3°C	Aboyne (Aberdeenshire)	21 Feb 2019
Northern Ireland	17.8°C	Bryansford (Co. Down)	13 Feb 1998
Minimum temperature			
England	-22.2°C	Scaleby (Cumbria) Ketton (Leicestershire)	19 Feb 1892 8 Feb 1895
Wales	-20.0°C	Welshpool (Powys)	2 Feb 1954
Scotland	-27.2°C	Braemar (Aberdeenshire)	11 Feb 1895
Northern Ireland	-15.6°C	Garvagh, Moneydig (Co. Londonderry)	20 Feb 1955

Country	Pressure	Location	Date
Maximum pressure			
Scotland	1052.9 hPa	Aberdeen Observatory	1 Feb 1902
Minimum pressure			
Republic of Ireland	942.3 hPa	Midleton (Co. Cork)	4 Feb 1951

Minimum temperature
-27.2°C
11 Feb 1895

Maximum temperature
18.3°C
21 Feb 2019

Maximum temperature
17.8°C
13 Feb 1998

Maximum pressure
1052.9 hPa
01 Feb 1902

Minimum temperature
-15.6°C
20 Feb 1955

Minimum temperature
-22.2°C
19 Feb 1892

Minimum temperature
-22.2°C
08 Feb 1895

Maximum temperature
20.8°C
26 Feb 2019

Minimum pressure
942.3 hPa
04 Feb 1951

Maximum temperature
21.2°C
26 Feb 2019

Minimum temperature
-20.0°C
02 Feb 1954

Aberdeen
Braemar Aboyne
Scaleby
Garvagh,
Moneydig
Bryansford
Porthmadog
Welshpool Ketton
Midleton
Kew
Gardens

The Weather in February 2023

Observation	Location	Date
Max. temperature 17.2°C	Pershore (Hereford & Worcester)	17 February
Min. temperature -8.5°C	Tulloch Bridge (Inverness-shire)	27 February
Most rainfall 45.6 mm	Cassley (Sutherland)	2 February
Highest gust 83 mph (72 Kt)	Baltasound No.2 (Shetland)	3 February
Greatest snow depth 1 cm	Fettercairn, Glensaugh No. 2 (Kincardineshire) and Oyne No. 2 (Aberdeenshire)	18 February

Trade winds

There are two trade-wind zones, north and south of the equator, with the north-east trades and the south-east trades, respectively, where the air converges on the low-pressure region at the equator. The direction and strength of these winds do remain relatively constant throughout the year, and were thus a reliable source of motive power for sailing ships

The start of the month saw strong westerly winds, with heavy rain disrupting train and ferry services in western Scotland. Cassley in Sutherland saw 45.6 mm of rain on the 2nd. Thereafter, February was remarkably dry, being dominated by a period of high pressure and predominantly anticyclonic. A cold easterly allowed temperatures to fall dramatically in the south, with South Newington in Oxfordshire recording a minimum of -8.4°C.

The middle of the month saw a series of depressions crossing the UK, bringing some rain along with very mild air. All four nations saw maxima in the mid-teens. The arrival of storm Otto on the 17th brought strong winds with gusts exceeding 60 mph (97 kph/84 knots) in many places across northern and eastern Scotland, north-east England and Yorkshire. Inverbervie in Kincardineshire saw gusts of 83 mph (134 kph/116 knots), and power outages affected as many as 26,000 in total. East Coast Main Line rail services were also severely disrupted by damage to overhead wires, and rural road networks impacted by trees being brought down.

Later in the month the high pressure zone was re-established, although winds were more northerly bringing temperatures down and widespread night-time frosts. No further adverse impacts were recorded, with the most notable phenomenon later in the month being an unusually prominent auroral display visible across much of Britain on the 26th.

Overall, dryness and mildness characterised the month. Central and southern England saw only 20 per cent of their average rainfall for February. Although rainfall was closer to normal further north and west, the UK as a whole received an average of only 43.4 mm of rain over the month, just 45 per cent of the February mean. Only north-west Scotland received more rain than average, and for the UK it was the driest February for 30 years. There was almost no snow, with the greatest depth recorded being 1 cm at Fettercairn on the 18th. Temperatures were well above average, with the provisional UK mean of 5.8°C some 1.7°C above the 1991–2020 average. This anomaly was particularly pronounced in Scotland and Northern Ireland, with mean minimums as high as 2.3°C above the 1991–2020 average.

Sunrise and Sunset 2024

Location	Date	Rise	Azimuth °	Set	Azimuth °
Belfast					
	1 Feb (Thu)	08:14	119	17:01	241
	11 Feb (Sun)	07:55	114	17:22	247
	21 Feb (Wed)	07:33	107	17:43	253
	29 Feb (Thu)	07:14	102	17:59	258
Cardiff					
	1 Feb (Thu)	07:52	117	17:01	243
	11 Feb (Sun)	07:35	112	17:20	248
	21 Feb (Wed)	07:16	106	17:38	254
	29 Feb (Thu)	06:59	101	17:52	259
Edinburgh					
	1 Feb (Thu)	08:08	120	16:45	240
	11 Feb (Sun)	07:48	115	17:07	246
	21 Feb (Wed)	07:25	108	17:29	252
	29 Feb (Thu)	07:05	103	17:46	258
London					
	1 Feb (Thu)	07:40	117	16:49	243
	11 Feb (Sun)	07:24	112	17:07	248
	21 Feb (Wed)	07:04	106	17:26	254
	29 Feb (Thu)	06:48	101	17:40	259

Note that all times are in Universal Time (UT), otherwise known as Greenwich Mean Time (GMT). These times do not take Summer Time (BST) into account.

Moonrise and Moonset 2024

Location	Date	Rise	Azimuth °	Set	Azimuth °
Belfast					
	1 Feb (Thu)	–	–	10:11	252
		23:38	104		
	11 Feb (Sun)	08:56	107	19:43	258
	21 Feb (Wed)	13:50	43	07:10	318
	29 Feb (Thu)	23:59	121	08:27	246
Cardiff					
	1 Feb (Thu)	–	–	10:06	253
		23:21	103		
	11 Feb (Sun)	08:38	106	19:35	259
	21 Feb (Wed)	14:00	47	06:37	314
	29 Feb (Thu)	23:35	118	08:25	247
Edinburgh					
	1 Feb (Thu)	–	–	09:57	252
		23:29	104		
	11 Feb (Sun)	08:48	108	19:29	258
	21 Feb (Wed)	13:28	41	07:10	321
	29 Feb (Thu)	23:53	122	08:12	245
London					
	1 Feb (Thu)	–	–	09:54	253
		23:09	103		
	11 Feb (Sun)	08:27	106	19:23	259
	21 Feb (Wed)	13:47	47	06:26	314
	29 Feb (Thu)	23:23	118	08:13	247

Note that all times are in Universal Time (UT), otherwise known as Greenwich Mean Time (GMT). These times do not take Summer Time (BST) into account.

Twilight Diagrams 2024

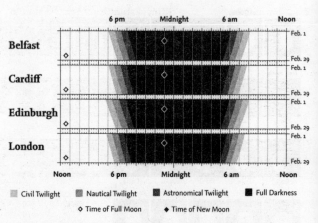

░ Civil Twilight	▓ Nautical Twilight	▓ Astronomical Twilight	█ Full Darkness
	◇ Time of Full Moon	◆ Time of New Moon	

The exact times of the Moon's major phases are shown on the diagrams opposite.

Front

A zone separating two air masses with different characteristics (typically, with different temperatures and/or humidities). Depressions normally show two fronts: a warm front (where warm air is advancing) and a cold front (where cold air is advancing). The latter normally move faster than warm fronts. When a cold front catches up with a warm front, the warm air is lifted away from the surface, giving a pool of warm air at altitude. The combined front is known as an occluded front. Depending on the exact conditions, occluded fronts may give long periods of overcast skies and persistent rain.

The Moon's Phases and Ages 2024

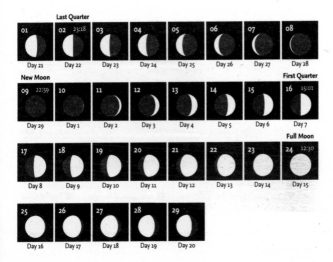

Last Quarter

| 01 Day 21 | 02 23:18 Day 22 | 03 Day 23 | 04 Day 24 | 05 Day 25 | 06 Day 26 | 07 Day 27 | 08 Day 28 |

New Moon | | | | | | | **First Quarter**

| 09 22:59 Day 29 | 10 Day 1 | 11 Day 2 | 12 Day 3 | 13 Day 4 | 14 Day 5 | 15 Day 6 | 16 15:01 Day 7 |

| | | | | | | | **Full Moon** |

| 17 Day 8 | 18 Day 9 | 19 Day 10 | 20 Day 11 | 21 Day 12 | 22 Day 13 | 23 Day 14 | 24 12:30 Day 15 |

| 25 Day 16 | 26 Day 17 | 27 Day 18 | 28 Day 19 | 29 Day 20 |

Cyclone
Technically, a name for any circulation of air around a low-pressure centre. (Depressions are also known as 'extratropical cyclones'.) The term is also used specifically for a tropical, revolving storm in the Indian Ocean, known as a 'hurricane' over the North Atlantic Ocean or eastern Pacific Ocean. The term 'typhoon' is used for systems in the western Pacific that affect northern Australia and Asia. The term 'tropical cyclones' applies to all such revolving systems.

February – In this month

1 February 1709 – Alexander Selkirk was rescued after being shipwrecked on a desert island. His story inspired Daniel Defoe's novel *Robinson Crusoe*.

1 February 1814 – The last Frost Fair to be held on the River Thames in London began. These events, which were held when the river in central London froze over for long enough, occurred roughly once in a generation during the Little Ice Age, from the seventeenth to nineteenth centuries. The one in 1814 lasted four days, and highlights included tradesmen and food stalls, dancing and nine-pin bowling. Several printing presses were set up on the ice, producing commemorative poems, and an elephant was led across the frozen river.

10 February 1947 – Severe winter conditions and a shortage of fuel led to major cuts in power supply in England and Wales. BBC television was shut down and didn't return until 11 March.

12 February 1928 – Heavy hailstorms resulted in eleven deaths in England.

18 February 2022 – Storm Eunice became one of the most powerful storms to hit the UK in decades, with the fastest wind gusts ever recorded in England blowing at 122 mph on the Isle of Wight. Three people died and widespread damage occurred, including the roof of the iconic O2 arena being partially blown away.

❄

22 February 2018 – Anticyclone Hartmut, also known as the Beast from the East, brought cold weather and heavy snowfall to large parts of the UK. Cold air from the Arctic reacted with the warmer water of the North Sea to cause the extraordinary snow depths.

❄

26 February 2019 – The highest UK temperature for the month of February was recorded in Porthmadog, Gwynedd, where it reached 20.8°C.

The railway at Dawlish in Devon

As anyone who has travelled by train to Plymouth, Cornwall and West Devon will know, at one stage the main line is extremely close to the sea. Indeed at Dawlish Warren and Dawlish itself, part of it runs along the sea wall, and from time to time the station at Dawlish has been closed because of waves breaking over the sea wall, putting trains at risk. The 'down' platform at Dawlish – closest to the sea – has been damaged several times, and the extremely strong storms of January and February 2014 delivered the final blow. After several storms in January, the wind and waves of 5 February 2014 brought utter devastation. The sea wall was breached, and the rails and sleepers of the railway left hanging in mid-air; the sea wall and the ballast from beneath the track had been swept away. What's more, a large section of the line was affected by an enormous cliff fall further along the coast at Teignmouth. All rail links to Cornwall and West Devon were cut. It being winter, the loss to the tourist industry was low, but fishing and other industries were hard hit by the loss of this link, which served as an economic lifeline for the region by transporting freight traffic.

There were immediate calls for the government to step in and provide an alternative route. One proposal was to undertake major improvements to the disused line through Okehampton and Tavistock that encircles the northern edge of Dartmoor. Opening that line to all rail traffic to Cornwall and the west of Devon would require major changes and investment to existing track, and would also involve major work to re-open the line running up the Teign valley, to connect the line to the main railway network. Another suggestion was to revive the 1930s proposal of the Dawlish Avoiding Line. Work on this line involved several tunnels through steep Devon hills and was started, but was halted by the outbreak of World War II, and never resumed.

In the event, sterling work by the combined effort of some 300 coastal and railway engineers and labourers brought the line back into service by April. The sea wall was reinstated – in effect, a new sea wall was constructed – some 25,000 tonnes were removed from the cliff fall at Teignmouth and hundreds of tonnes of debris carted away at Dawlish itself. Nearly a kilometre of track and ballast was re-laid, and many kilometres of new electrical cables were installed.

Nevertheless, one estimate of the cost to the industries (including the tourist and fishing industries) of west Devon and Cornwall incurred by the disruption amounted to between £60 million and £1.2 billion. Conversation surrounding the long-term sustainability of the railway line in the face of worsening weather patterns as a result of climate change has not abated. Part of the alternative route, from Exeter to Okehampton, was reopened as a branch line in 2021. In 2019, the whole northern route was listed by the Campaign for Better Transport as a 'priority 1' candidate for reopening.

The condition of the railway at Dawlish after the disastrous storm of 5 February 2014. The sea wall and railway ballast have been washed away.

March

Introduction

March has traditionally been viewed as a transitional month between winter and spring. It has come to be associated with the saying 'In like a lion, out like a lamb'. Regrettably for this saying, although the weather may well be changeable, there is no distinct pattern that applies every year. The month may well begin with a settled period, with little in the way of winds and dramatic events, and then deteriorate. Interestingly, there is often a sharp change in the weather after the first week of March. The frequency of westerly weather – that is, the progression of depressions across the country from the west – drops suddenly, and very few are recorded. The number of such depressions and sequences of weather is less than at any other time of the year, except for late April and early May. In general, the increasing strength of heating from the Sun begins to take effect from the beginning of the month. Although there may still be spells of very cold weather, with a lot of cloud, the increased warmth is readily noticed.

Another tradition, strongly held by some, particularly by mariners, is that March experiences 'equinoctial gales'. (The equinox is on 20 March in 2024.) This idea is, however, not supported by the evidence. Gales are actually most common around the winter solstice (21 December in 2024), and least frequent around the summer solstice (20 June in 2024). There may be a perception that gales are most frequent around the autumnal equinox (22 September in 2024), but this is probably a result of the deterioration of the weather from the fine, long days of summer to the shorter, less settled days of autumn. When the Sun is well north of the equator, its warmth generally causes the Azores High to strengthen and extend its influence as ridges of high pressure. These may reach so far as to cover the British Isles and most of western Europe. Sometimes the high-pressure region extends well to the west, giving rise to what is known as the Bermuda High. Depressions, and their accompanying cloud and winds, tend to follow the Polar Front jet stream, which is diverted towards the north. The low-pressure systems, with their clouds, rain and winds, pass north of the British Isles, which thus remains under the influence of the settled, anticyclonic weather within the extended Azores High. With the change of season from summer to autumn, the Sun starts to move south. It moves fastest towards the equator and south of it in September, usually causing the high-pressure area over the Azores to decline. When this happens, depressions are no longer displaced northwards and they, and their accompanying winds, thus cross the British Isles more directly. It would seem that mariners' observations of such windier weather in September have been taken in a more general sense to suggest that gales occur at both equinoxes (in March and September). There is no evidence whatsoever for increased windiness at either equinox, so this is yet another instance of incorrect weather lore.

M

Azores High

A more-or-less permanent high-pressure system in the North Atlantic, generally centred approximately over the islands of the Azores, or closer to Iberia (Portugal and Spain). It arises from air that has risen at the equator that descends at the sub-tropical high-pressure zones.

Weather Extremes in March

Country	Temp.	Location	Date
Maximum temperature			
England	25.6°C	Mepal (Cambridgeshire)	29 Mar 1968
Wales	23.9°C	Prestatyn (Denbighshire) Ceinws (Powys)	29 Mar 1965
Scotland	23.6°C	Aboyne (Aberdeenshire)	27 Mar 2012
Northern Ireland	21.8°C	Armagh (Co. Armagh)	29 Mar 1965
Minimum temperature			
England	-21.1°C	Houghall (Co. Durham)	4 Mar 1947
Wales	-21.7°C	Corwen (Denbighshire)	3 Mar 1965
Scotland	-22.8°C	Logie Coldstone (Aberdeenshire)	14 Mar 1958
Northern Ireland	-14.8°C	Katesbridge (Co. Down)	2 Mar 2001

Country	Pressure	Location	Date
Maximum pressure			
Republic of Ireland	1051.3 hPa	Malin Head (Co Donegal)	29 Mar 2020
Minimum pressure			
Scotland	946.2 hPa	Wick (Caithness)	9 Mar 1876

Minimum
temperature
-22.8°C
14 Mar 1958

Minimum
pressure
946.2 hPa
09 Mar 1876

Maximum
temperature
23.6°C
27 Mar 2012

Wick

Logie Coldstone ▲▲ Aboyne

Maximum
pressure
1051.3 hPa
29 Mar 2020

Minimum
temperature
-21.1°C
04 Mar 1947

Malin Head ▲

Houghall

Armagh ▲ Katesbridge

Maximum
temperature
23.9°C
29 Mar 1965

Maximum
temperature
21.8°C
29 Mar 1965

Prestatyn
Corwen

Ceinws ▲

Mepal

Minimum
temperature
4.8°C
2 Mar 2001

Maximum
temperature
23.9°C
29 Mar 1965

Maximum
temperature
25.6°C
29 Mar 1968

Minimum
temperature
-21.7°C
03 Mar 1965

M

The Weather in March 2023

Observation	Location	Date
Max. temperature 17.8°C	Santon Downham (Suffolk)	30 March
Min. temperature -16.0°C	Altnaharra No 2 (Sutherland)	9 March
Most rainfall 118.6 mm	Honister Pass (Cumbria)	12 March
Most sunshine 11.9 hrs	Bishopton (Renfrewshire)	27 March
Greatest snow depth 32 cm	Buxton, Derbyshire	10 March

With high pressure continuing, March began as February had ended: cold and dry. Some snow arrived on a surge of Arctic air on the 6th and 7th, and from the 8th onwards things became more unsettled as moist, milder air intruded from the south. The boundary between these two systems witnessed more snow, with disruption reported at Bristol airport and schools closed in large parts of Scotland. In Derbyshire a major incident was declared on the 9th, and mountain rescue teams were deployed to help motorists trapped between Buxton and Ashbourne. The 9th also saw the lowest minimum, as temperatures at Altnaharra in Sutherland hit -16°C. The following morning saw 32 cm of snow lying at Buxton.

After persisting for several days, the cold northern air was finally displaced by the 16th as the high pressure drifted away towards Greenland. In its place a series of Atlantic depressions drove across the UK. The middle of the month was therefore milder though unsettled, with westerly and south-westerly winds and maximum temperatures exceeding 16°C in all four nations. Several bands of rain in quick succession saw volumes in excess of 50 mm at numerous sites on the 12th. Honister Pass in Cumbria (a notoriously wet site) received more than twice that, with 118.6 mm. More heavy rain on the 18th and high spring tides on the 22nd brought widespread flooding everywhere from South Yorkshire to Cornwall.

Thereafter followed a few days of finer weather, with 11.9 hours of sunshine recorded in Bishopton, Renfrewshire on the 27th, but the very end of the month became unsettled again. A low pressure zone passed close to the south of the UK, bringing strong winds. In Cornwall there were reports of power outages and trees blown down, but for most of the UK the end of the month was dull and wet.

Overall mean temperatures were close to average in many places, although northern Scotland was colder. The UK mean temperature of 5.7°C was equal to the 1991–2020 average. Rainfall was a different matter; only north-western Scotland received less than normal, with the UK overall 155 per cent wetter than average. It was the sixth wettest March since 1836, and some places in the south of England and Wales saw only half the average amount of sunshine, making it the dullest March since 1910 for some counties.

Rain Shadow

An area to the leeward of high ground, whether hills or mountains, often experiences less rainfall than neighbouring areas or than expected. This rain-shadow effect occurs because as air rises over the higher ground (usually to the west) there is increased rainfall, leaving less moisture to fall on any areas to the leeward of the hills.

Sunrise and Sunset 2024

Location	Date	Rise	Azimuth °	Set	Azimuth °
Belfast					
	1 Mar (Fri)	07:12	102	18:01	259
	11 Mar (Mon)	06:48	95	18:21	265
	21 Mar (Thu)	06:23	88	18:40	272
	31 Mar (Sun)	05:57	81	18:59	279
Cardiff					
	1 Mar (Fri)	06:57	101	17:54	259
	11 Mar (Mon)	06:35	95	18:11	266
	21 Mar (Thu)	06:12	88	18:28	272
	31 Mar (Sun)	05:49	82	18:45	278
Edinburgh					
	1 Mar (Fri)	07:03	102	17:48	258
	11 Mar (Mon)	06:37	95	18:09	265
	21 Mar (Thu)	06:11	88	18:29	272
	31 Mar (Sun)	05:45	81	18:50	279
London					
	1 Mar (Fri)	06:45	101	17:42	259
	11 Mar (Mon)	06:23	95	17:59	266
	21 Mar (Thu)	06:00	88	18:16	272
	31 Mar (Sun)	05:38	82	18:33	278

Note that all times are in Universal Time (UT), otherwise known as Greenwich Mean Time (GMT). These times do not take Summer Time (BST) into account.

Moonrise and Moonset 2024

Location	Date	Rise	Azimuth °	Set	Azimuth °
Belfast					
	1 Mar (Fri)	–	–	08:37	236
		23:59	121		
	11 Mar (Mon)	07:20	89	20:16	278
	21 Mar (Thu)	14:15	57	05:53	306
	31 Mar (Sun)	01:51	142	04:51	78
Cardiff					
	1 Mar (Fri)	–	–	08:39	239
		23:35	118		
	11 Mar (Mon)	07:10	89	20:01	277
	21 Mar (Thu)	14:17	60	05:28	303
	31 Mar (Sun)	01:14	137	08:00	321
Edinburgh					
	1 Mar (Fri)	–	–	08:19	235
		23:53	122		
	11 Mar (Mon)	07:09	89	20:06	278
	21 Mar (Thu)	13:57	56	05:49	307
	31 Mar (Sun)	01:53	145	07:21	214
London					
	1 Mar (Fri)	–	–	08:27	239
		23:23	118		
	11 Mar (Mon)	06:58	89	19:49	277
	21 Mar (Thu)	13:47	47	06:26	314
	31 Mar (Sun)	01:02	138	07:47	221

Note that all times are in Universal Time (UT), otherwise known as Greenwich Mean Time (GMT). These times do not take Summer Time (BST) into account.

Twilight Diagrams 2024

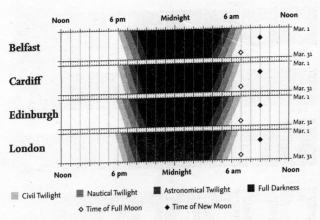

The exact times of the Moon's major phases are shown on the diagrams opposite.

Air mass
A large volume of air that has uniform properties (particularly temperature and humidity) throughout. Air masses arise when air stagnates over a particular area for a long time. These areas are known as 'source regions' and are generally the semi-permanent high-pressure zones, which are the sub-tropical and polar anticyclones. The primary classification is based on temperature, giving Arctic (A), polar (P) and tropical (T) air.

The Moon's Phases and Ages 2024

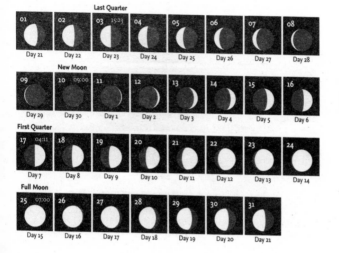

Last Quarter							
01	02	03 15:23	04	05	06	07	08
Day 21	Day 22	Day 23	Day 24	Day 25	Day 26	Day 27	Day 28

New Moon							
09	10 09:00	11	12	13	14	15	16
Day 29	Day 30	Day 1	Day 2	Day 3	Day 4	Day 5	Day 6

First Quarter							
17 04:11	18	19	20	21	22	23	24
Day 7	Day 8	Day 9	Day 10	Day 11	Day 12	Day 13	Day 14

Full Moon						
25 07:00	26	27	28	29	30	31
Day 15	Day 16	Day 17	Day 18	Day 19	Day 20	Day 21

M

Lapse rate
The change in a property with increasing altitude. In meteorology, this is usually the change in temperature. In the troposphere (the lowest layer of the atmosphere), this is a decrease in temperature with an increase in height. This is defined as a positive lapse rate. In the stratosphere (the next higher layer) there is an overall increase in temperature with height, giving a negative lapse rate.

March – In this month

2 March 1963 – A family on Dartmoor was finally rescued after being stranded for 65 days in a remote farmhouse among 8-metre drifts of snow.

2 March 1948 – A Sabena Douglas DC-3 aircraft on a flight from Brussels to London missed the runway in thick fog and crashed. Although three passengers were rescued, all three crew and 16 of the passengers died in the resulting fire. It was the first (and worst) accident at London's Heathrow airport.

10 March 1891 – A violent, heavy snowstorm in the West Country introduced a new word to the English language. It was the very first event to be described by the word 'blizzard'.

12 March 1744 – A strong gale in the English Channel prevented a French invasion fleet, inspired by James III's attempt to claim the throne of England and Scotland, from leaving Dunkirk.

15 March 1789 – The Adventure sank off South Shields, with the loss of all its crew. This prompted a competition to design the first purpose-built lifeboat, with the winning entry bult by Henry Greathead and called the Original.

16 March 1947 – The 'Great Suck' began. Thawing of the enormous snowfall led to breaches in the dykes in the Fens and widespread flooding. Local authorities, the fire brigade and the army began to use pumps to reduce the flooding. Some land was not freed of lying water until June.

19 March 1969 – During strong winds, the Emley Moor television mast collapsed under the weight of the ice on the tower and guy wires. At the time of its construction in 1966, it was one of the tallest structures in the world at 385 metres.

21 March 1748 – John Newton encountered a violent Atlantic storm during a passage home. Twenty years later it inspired him (together with the poet William Cowper) to write 'Amazing Grace', still the most popular hymn in Britain.

27 March 1980 – A ferocious windstorm in the North Sea caused the Alexander Kielland accommodation platform to collapse off the Scottish coast. Some 123 workers were drowned and an enormous international rescue effort was launched, involving nearly 50 vessels, 27 helicopters and two planes. Eventually, 89 men were rescued, proving the value of the use of rescue helicopters.

Four Inns Walk 1964

The Four Inns Walk was a hill-walking event organised by the Scouting Association. It was mainly carried out over the high moorland of the northern Peak District in Derbyshire, although starting at Holmbridge in Yorkshire and with a short diversion into Cheshire. The route was 65 km long, taking in Snake Pass Inn, the Nag's Head and the Cat and Fiddle, as well as the site of a former inn called the Isle of Sky. The event began in 1957, originally for Rover Scouts (those over the age for Boy Scouts, usually regarded as 14–15 years old), and undertaken in teams of three or four participants.

The 1964 event, held in mid-March, was particularly disastrous. Of 240 who started the event, only 22 managed to finish. No fewer than three participants died. The area's rapidly changing weather conditions are well documented, and the beautiful landscape can turn into one of bleak danger very quickly. Unfortunately, on this occasion the weather forecast was incorrect, predicting favourable conditions. The weather suddenly deteriorated, with high winds, driving, heavy rain, and low temperatures.

The first to die was Gordon Stuart Withers, the youngest (aged 19) of a team from Huddersfield, who got into trouble above Snake Pass. He was rescued by members of another

The trig point at Higher Shelf Stones, overlooking Snake Pass.

Part of the mountain pass, known as the Snake Pass, in Derbyshire, where the road crosses the Pennines. The area is notorious for its rapid changes of weather.

team and taken to hospital, where he subsequently succumbed to hypothermia. The other two, Michael Welby and John Butterfield, were older (ages 21 and 24) and from another team. Their group encountered bad weather in the same area. A search party was sent out and a third member of their team was rescued but he was unable to give any information about the location of the other members. With the weather worsening, the search had to be called off overnight, resuming on Sunday morning. The first body was recovered on Monday, and on Tuesday morning the second body was recovered.

The tragedy prompted the formation of a dedicated Peak District Mountain Rescue Organisation. Meanwhile, because of difficulties in organising the Four Inns Walk, it has now been replaced in the calendar by two related but shorter hill-walking events, known as the Kinder Extreme and the Kinder Challenge. Both events start and finish at Chapel-en-le-Frith in the Peak District. The Kinder Challenge is about 16 km, and the Kinder Extreme about 60 km, including part of the Kinder Challenge route.

April

Introduction

April is traditionally associated with a changeable month and the general view is that it is accompanied by 'April showers'. Although it is certainly changeable, with the transition from winter to the warmer conditions of summer, in recent years the showers have been less frequent. Certainly, there are normally periods when northerly winds (in particular) bring showers and these may be squally, and can sometimes even turn into thunderstorms and be accompanied by hail. Generally nowadays, the showers are infrequent, and it is one of the quietest times of the year. April is actually one of the driest months for most of the country, followed in sequence by March, June, May and February.

The temperature of the sea surrounding the British Isles remains low, while the land begins to warm up. Sea breezes are common and they are often accompanied by sea fog. This is particularly frequent on the east coast of Britain. Sea fogs often form over the cold North Sea and then are carried inland by the sea breezes that build up during the day. In Scotland, in particular, the 'haar', as it is known, often moves in during the afternoon. It is often particularly noticeable when it invades the Firth of Forth and hides the bottom of the Forth Bridges from view. A similar phenomenon occurs anywhere along the east coast of England, particularly in Northumbria, but is found as far south as East Anglia. The same sort of sea fog does occur on the western coasts of Britain, but is less common, because the sea surface temperature of the Atlantic water tends to be higher than that of the North Sea, so sea fog is less likely to form.

A

Spring and early summer season – 1 April to 17 June

The weather during this season is very changeable. Indeed, it is the most changeable of the year. There are occasional outbreaks of northerly winds, which tend to produce heavy squally showers, often turning into thunderstorms with lightning and even hail. Initially, high pressure and dry air over Continental Europe often extends west over the British Isles, but this high pressure and its accompanying dry, continental air tend to collapse in early summer and be replaced by moist maritime air as depressions move across the country from the west, on tracks towards the Baltic or slightly farther north, towards Scandinavia.

Weather Extremes in April

Country	Temp.	Location	Date
Maximum temperature			
England	29.4°C	Camden Square (London)	16 Apr 1949
Wales	26.2°C	Gogerddan (Ceredigion)	16 Apr 2003
Scotland	27.2°C	Inverailort (Highland)	17 Apr 2003
Northern Ireland	24.5°C	Boom Hall (Co. Londonderry)	26 Apr 1984
Minimum temperature			
England	-15.0°C	Newton Rigg (Cumbria)	2 Apr 1917
Wales	-11.2°C	Corwen (Denbighshire)	11 Apr 1978
Scotland	-15.4°C	Eskdalemuir (Dumfriesshire)	2 Apr 1917
Northern Ireland	-8.5°C	Killylane (Co. Antrim)	10 Apr 1998

Country	Pressure	Location	Date
Maximum pressure			
Scotland	1044.5 hPa	Eskdalemuir (Dumfrieshire)	11 Apr 1938
Minimum pressure			
Republic of Ireland	952.9 hPa	Malin Head (Co. Donegal)	1 Apr 1948

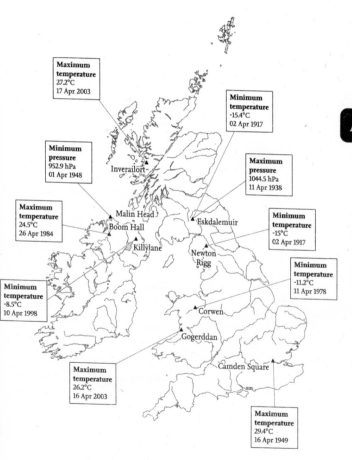

Maximum temperature
27.2°C
17 Apr 2003

Minimum temperature
-15.4°C
02 Apr 1917

Minimum pressure
952.9 hPa
01 Apr 1948

Maximum pressure
1044.5 hPa
11 Apr 1938

Inverailort

Maximum temperature
24.5°C
26 Apr 1984

Minimum temperature
-15°C
02 Apr 1917

Malin Head
Boom Hall
Killylane
Eskdalemuir
Newton Rigg

Minimum temperature
-11.2°C
11 Apr 1978

Minimum temperature
-8.5°C
10 Apr 1998

Corwen
Gogerddan

Camden Square

Maximum temperature
26.2°C
16 Apr 2003

Maximum temperature
29.4°C
16 Apr 1949

A

The Weather in April 2023

Observation	Location	Date
Max. temperature 21.2°C	Kinlochewe (Ross & Cromarty)	17 April
Min. temperature -7.4°C	Tulloch Bridge (Inverness-shire)	25 April
	Loch Glascarnoch (Ross & Cromarty)	26 April
Most rainfall 54.6 mm	Seathwaite (Cumbria)	11 April
Most sunshine 14.3 hrs	Loch of Hundland (Orkney)	21 April
Highest gust 96 mph (83 Kt)	Needles Old Battery (Isle of Wight)	12 April

April saw little of note, with weather mostly staying within the parameters of normal for spring. The jet stream was displaced southwards, bringing stronger winds across the south of the UK, with calmer weather further north. However, high pressure was dominant to the east or north of the country for most of the month, meaning any weather systems coming off the Atlantic were fairly weak. Severe weather warnings were sparse and the subdued weather caused relatively low maximum temperatures. Frosts were more common than usual in sheltered areas, and Benson in Oxfordshire saw an overnight low of -5.6°C on the 4th.

The exception to this picture occurred around the middle of the month. On the 11th and 12th a pair of low pressure centres interacted over the UK, bringing strong winds to Wales and southern England. Cairngorm Summit registered gusts of 91 mph (148 kph/127 knots) on the 11th, and the Needles on the Isle of Wight recorded 96 mph (153 kph/83 knots) on the 12th. There were reports of HGVs overturning on the M6 in Cumbria and on the M62 on the border of Greater Manchester and West Yorkshire. Rainfall was also high, with Seathwaite in Cumbria (holder of the UK 24-hour maximum rainfall) recording 54.6 mm. The system was named by MeteoFrance as storm Noa.

The high pressure re-established itself from then on, and southerly winds brought milder temperatures. Kinlochewe in Ross and Cromarty reached 21.2°C on the 17th. Although temperatures fell around the 24th and 25th, the very end of the month saw widespread temperatures in the high teens and low 20s again. The highest minimum of the month was recorded at Castlederg, Tyrone on the 30th, at 11.7°C.

In sum, April was unsettled without being extreme. Temperatures and rainfall fluctuated over the month and between regions, but both averaged around normal overall. Northern Ireland was warmest relative to average, while most of England saw averages slightly below normal. Across the UK, the provisional mean temperature for the month of 7.8°C was 0.1°C below average. Most parts of Scotland were drier than average, but southern and eastern England saw more rainfall than usual. The UK as a whole received 97 per cent of the average rain for the month, and 102 per cent of average sunshine.

Humidity

A measure of the quantity of water vapour in the air. It generally increases with an increase in temperature.

Sunrise and Sunset 2024

Location	Date	Rise	Azimuth °	Set	Azimuth °
Belfast					
	1 Apr (Mon)	05:55	81	19:01	280
	11 Apr (Thu)	05:30	74	19:20	286
	21 Apr (Sun)	05:06	68	19:40	293
	30 Apr (Tue)	04:46	62	19:57	298
Cardiff					
	1 Apr (Mon)	05:47	81	18:47	279
	11 Apr (Thu)	05:25	75	19:03	285
	21 Apr (Sun)	05:04	69	19:20	291
	30 Apr (Tue)	04:46	64	19:35	296
Edinburgh					
	1 Apr (Mon)	05:42	80	18:52	280
	11 Apr (Thu)	05:16	73	19:12	287
	21 Apr (Sun)	04:52	67	19:32	294
	30 Apr (Tue)	04:30	61	19:51	299
London					
	1 Apr (Mon)	05:35	81	18:35	279
	11 Apr (Thu)	05:13	75	18:52	285
	21 Apr (Sun)	04:52	69	19:09	291
	30 Apr (Tue)	04:34	64	19:24	296

Note that all times are in Universal Time (UT), otherwise known as Greenwich Mean Time (GMT).

Moonrise and Moonset 2024

Location	Date	Rise	Azimuth °	Set	Azimuth °
Belfast					
	1 Apr (Mon)	–	–	08:37	236
		23:59	121		
	11 Apr (Thu)	07:20	89	20:16	278
	21 Apr (Sun)	14:15	57	05:53	306
	30 Apr (Tue)	01:51	142	04:51	78
Cardiff					
	1 Apr (Mon)	–	–	08:39	239
		23:35	118		
	11 Apr (Thu)	07:10	89	20:01	277
	21 Apr (Sun)	14:17	60	05:28	303
	30 Apr (Tue)	01:14	137	08:00	321
Edinburgh					
	1 Apr (Mon)	03:06	149	08:06	211
	11 Apr (Thu)	05:54	47	–	–
				00:02	320
	21 Apr (Sun)	16:50	94	14:24	272
	30 Apr (Tue)	02:34	143	08:28	218
London					
	1 Apr (Mon)	02:11	140	08:37	219
	11 Apr (Thu)	06:07	52	23:17	313
	21 Apr (Sun)	16:36	93	04:11	272
	30 Apr (Tue)	01:46	137	08:51	225

Note that all times are in Universal Time (UT), otherwise known as Greenwich Mean Time (GMT).

Twilight Diagrams 2024

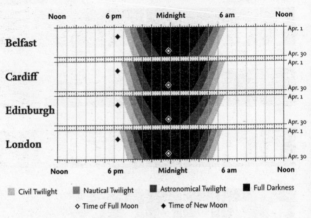

The exact times of the Moon's major phases are shown on the diagrams opposite.

Sub-tropical highs
Semi-permanent areas in both hemispheres around the latitudes of approximately 30° north and south, where air that has risen at the equator descends back to the surface, becoming heated and dry as it does so. They form the descending limbs of the Hadley cells – the cells closest to the equator.

The Moon's Phases and Ages 2024

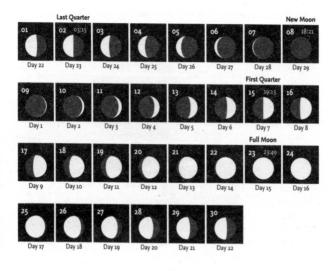

Depression

A low-pressure area. (Often called a 'storm' in North American usage.) Winds circulate around a low-pressure centre in an anticlockwise direction in the northern hemisphere. (Clockwise in the southern hemisphere.) Away from the surface, and the friction that it causes, winds flow along the isobars. Depressions generally move across the globe from west to east, although under certain conditions they may linger over an area or even (rarely) move towards the west.

April – In this month

1 April 1875 – The first newspaper weather map was published in *The Times*. The map showed the situation on the previous day, March 31. It was prepared by Francis Galton, a fierce critic of Robert FitzRoy's empirical methods of preparing weather forecasts.

3 April 2001 – Several months of heavy rainfall, the heaviest since records began in 1766, caused some 160 metres of the chalk cliffs at Beachy Head in East Sussex to collapse into the sea.

4 April 1635 – Extreme hail fell at Castletown, County Offaly. The stones were 4 inches (about 100 mm) in circumference.

5 April 1911 – The largest tree in England was blown down by a snowstorm in Oxford. The tree was the Huntingdon wych elm in the courtyard at Magdalen College. It had a height of 43 metres, a circumference of over 8 metres and was over 400 years old.

5 April 1635 – Twelve-year-old Peter Chivers invented the sport of windsurfing in a gentle breeze at Hayling Island in southern Hampshire.

10 April 1815 – The cataclysmic climax of the eruption of Mount Tambora had such widespread effects on the atmosphere that the whole world was cooled, leading to the failure of harvests the following year. 1816 became known as 'The Year without a Summer'.

10 April 1736 – The first swallow arrived at Stratton Strawless in Norfolk. The fact was noted down (for the first time ever) by Robert Marsham, who thus invented the science of phenology (the study of the occurrence of natural phenomena).

20 April 1802 – Dorothy Wordsworth noted in her diary how during a walk on 15 April 1802, she and her brother noticed daffodils tossing in the wind. Her journal later inspired her brother William to write his most famous poem 'I Wandered Lonely as a Cloud'.

21 April 1850 – Dr George Merryweather of Whitby claimed that a storm off the Yorkshire coast that day was accurately predicted by his 'jury of philosophical counsellors'. His 'jury' consisted of twelve leeches, and Merryweather claimed that their behaviour accurately predicted the weather.

23 April 1471 – The Battle of Barnet during the Wars of the Roses took place in such heavy mist that the soldiers of both sides could not distinguish whom they were fighting. The Earl of Warwick's men fought with those of the Earl of Oxford, thinking that they were engaging with King Edward's men. The resultant chaos marked the beginning of the end of the Wars of the Roses.

Below: *The very first weather forecast, prepared by Robert FitzRoy, and published in* The Times *of 1 August 1861.*

THE WEATHER.

METEOROLOGICAL REPORTS.

Wednesday, July 31, 8 to 9 a.m.	B.	E.	M.	D.	F.	C.	I.	S.
Nairn..	29·54	57	56	W.S.W.	6	9	o.	3
Aberdeen	29·60	59	54	S.S.W.	5	1	b.	3
Leith	29·70	61	55	W.	3	5	c.	2
Berwick	29·69	59	55	W.S.W.	4	4	c.	2
Ardrossan ..	29·73	57	55	W.	5	4	c.	5
Portrush	29·72	57	54	S.W.	2	2	b.	2
Shields	29·90	59	54	W.S.W.	4	5	o.	3
Galway	29·63	65	62	W.	5	4	c.	4
Scarborough ..	29·86	59	56	W.	3	6	c.	2
Liverpool	29·91	61	56	S.W.	2	8	c.	3
Valentia	29·37	62	60	S.W.	2	5	o.	3
Queenstown ..	29·58	61	59	W.	3	5	c.	2
Yarmouth.. ..	30·05	61	59	W.	5	2	c.	3
London	30·02	62	56	S.W.	3	2	b.	—
Dover..	30·04	70	61	S.W.	3	7	o.	2
Portsmouth ..	30·01	61	59	W.	3	6	o.	2
Portland	30·03	63	59	S.W.	3	2	c.	3
Plymouth.. ..	30·00	62	59	W.	5	1	b.	4
Penzance	30·04	61	60	S.W.	2	6	c.	3
Copenhagen ..	29·94	64	—	W.S.W.	2	6	c.	3
Helder	29·99	63	—	W.S.W.	6	5	c.	3
Brest	30·09	60	—	S.W.	2	6	o.	5
Bayonne	30·13	68	—	—	—	9	m.	5
Lisbon	30·13	70	—	N.N.W.	4	3	b.	2

General weather probable during next two days in the—
North—Moderate westerly wind ; fine.
West—Moderate south-westerly ; fine.
South—Fresh westerly ; fine.

Explanation.
B. Barometer, corrected and reduced to 32° at mean sea level ; each 10 feet of vertical rise causing about one-hundredth of an inch diminution, and each 10° above 32° causing nearly three-hundredths increase. E. Exposed thermometer in shade. M. Moistened bulb (for evaporation and dew-point). D. Direction of wind (true—two points *left* of magnetic). F. Force (1 to 12—estimated). C. Cloud (1 to 9). I. Initials :—b., blue sky ; c., clouds (detached) ; f., fog ; h., hail ; l, lightning ; m., misty (hazy) ; o., overcast (dull) ; r., rain ; s., snow ; t., thunder. S. Sea disturbance (1 to 9).

Next page: *The first weather map, prepared by Francis Galton, and published in* The Times *of 1 April 1875, 14 years after FitzRoy's forecast.*

WEATHER CHART, MARCH 31, 1875.

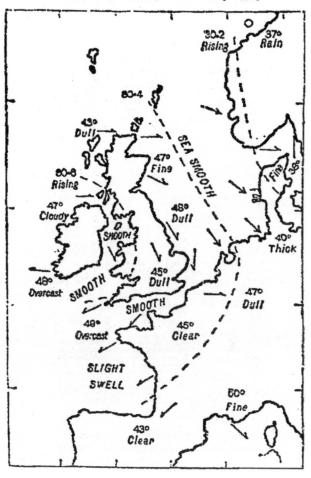

A

Phenology

On 10 April 1736, Robert Marsham recorded the first swallow to arrive at Stratton Strawless in Norfolk. This was the first time ever that anyone had recorded details of the date of an event in the natural world.

In effect, Robert Marsham invented the science that has since become known as 'phenology'. This unusual word is actually derived from two Greek words: one φαινω (*phaino*), meaning 'show', 'appear' and 'bring to light' and the more familiar λογος (*logos*), with the meanings of 'study' and 'reasoning'. The term was first employed in 1842 and 1843, and is used to describe the study of the timings of cyclical or seasonal biological events. In practice, this means the measuring and recording of the first occurrence of such events each year – the first sighting of swallows and butterflies, and the dates on which particular flowers come into bloom, or leaves emerge.

The date at which migratory butterflies are first seen – as with this small tortoiseshell – are significant in the study of phenology.

Other examples of phenomena studied include the dates of leaf fall from deciduous trees in autumn, and the dates of egg-laying in birds.

This information is of particular relevance to the study of climate change because it provides information on long-term alterations in the natural world. These dates are generally related to temperature (technically, they are known as a 'proxy' of temperature) and so recording their pattern helps show climate trends over time. Comparing these measurements to today's provides a key indicator of the effects of global warming.

Phenology is also involved in the study of how factors such as elevation play a part in the occurrence of certain plants and animals; not only do the species present vary by altitude, but also the times at which given behaviours occur. Once again, changes in the timings of these cycles can be a useful marker of how climate change is affecting particular locations.

While some behaviours might be directly temperature-dependent, others are secondarily so. For instance, the migration of birds is linked to the availability of food resources in particular areas, such as berries or insects. These are often dependent on plants for their own sources of food, and so a chain of dependence is created. A change in the seasonal timings of one link in the chain could affect the other links too. It is not only changes in the timing of the natural cycle of events that may be noted. For example, there are specific qualitative changes in the development of honey-bee colonies that are related to the temperature that prevails in their environment, so recording the types of bees and their behaviour also may be of interest. Similarly, another source of phenological information is records of wine production. These have been used to create a record of seasonal temperatures over the last 500 years based on the amount and quality of different varieties of wine produced each year. Such records show, for example, that during a warm phase, wine production once existed much farther north in Britain than is currently possible. Now, climate change is enabling a recurrence of this, with wine production recently having resumed, with some great success, in Britain.

May

Introduction

The month of May has traditionally been associated with the beginning of summer, with innumerable celebrations taking place on 1 May. Many of these events are very ancient festivals to mark the first day of summer.

In England, festivities on 1 May frequently included maypole dancing (often thought to be a form of fertility rite), Morris dancing, and sometimes the appearance of a Green Man. In Scotland, May Day celebrations, which have taken place for centuries, were formerly closely associated with the ancient rite of Beltane. The latter was one of the four traditional Gaelic festivals, celebrated approximately halfway between the spring equinox and the summer solstice. Over the years it had become established as May 1. In Wales, celebrations were, like those in Scotland, originally a continuation of the ancient Beltane festivities.

Although it is true that May, more than any other month, experiences more periods of high-pressure (anticyclonic) conditions, with long spells of fine, settled weather, records show that May has also experienced its fair share of extreme weather, with major snowfalls as well as extreme heat. Bank holidays for England had been established in 1871 under William Gladstone as Prime Minister. The original intention was for a bank holiday to occur at Whitsun. But Whitsun is a religious date, seven weeks after Easter and, although it often fell in May, could fall on any day between 11 May and 14 June. With such a range of dates, the weather could be extremely variable. Twenty years after the introduction of bank holidays, on the Whitsun Bank Holiday on 18 May 1891, there was a major snowfall that blanketed East Anglia and the temperature fell to -8.6°C in the notorious Rickmansworth frost hollow. Snow is not entirely uncommon this late in the year. Snow in Yorkshire was particularly deep on 17 May 1935. In mid-May 1955, the worst snowstorm for 60 years affected a large area covering Birmingham, the Cotswolds and the Chiltern Hills. On 19 May 1996, heavy snow on Dartmoor caused the annual Ten Tors adventure training weekend, run by the army, to be abandoned.

Other May weather has been very changeable. For example, that Ten Tors weekend had temperatures of 26°C in 1997, and was abandoned again in 2007, when a young girl was swept away by a brook, swollen from a depth of 1 metre to 5 metres by heavy rain. The wettest May day ever was 7 May 1881, when no less than 172.2 mm of rain fell in Cumbria. Extreme thunderstorms occurred on Derby Day, 31 May 1911. In May 1923, Scotland Yard complained that fog in London was hampering traffic and preventing the detection of crime.

M

Synoptic
The term 'synoptic' is used extensively in meteorology to indicate that the data used in preparing a chart (for example) were all obtained at the same time and thus show the state of the atmosphere at a particular moment.

Weather Extremes in May

Country	Temp.	Location	Date
Maximum temperature			
England	32.8°C	Camden Square (London)	22 May 1922
		Horsham (West Sussex)	29 May 1944
		Tunbridge Wells (Kent)	29 May 1944
		Regent's Park (London)	29 May 1944
Wales	30.6°C	Newport (Monmouthshire)	29 May 1944
Scotland	30.9°C	Inverailort (Highland)	25 May 2012
Northern Ireland	28.3°C	Lisburn (Co. Antrim)	31 May 1922
Minimum temperature			
England	-9.4°C	Lynford (Norfolk)	4 May 1941 / 11 May 1941
Wales	-6.2°C	St Harmon (Powys)	14 May 2020
Scotland	-8.8°C	Braemar (Aberdeenshire)	1 May 1927
Northern Ireland	-6.5°C	Moydamlaght (Co. Londonderry)	7 May 1982

Country	Pressure	Location	Date
Maximum pressure			
Republic of Ireland	1043.0 hPA	Sherkin Island (Co. Cork) / Valentia Obsy. (Co. Kerry)	12 May 2012
Minimum pressure			
England	968.0 hPa	Sealand (Cheshire)	8 May 1943

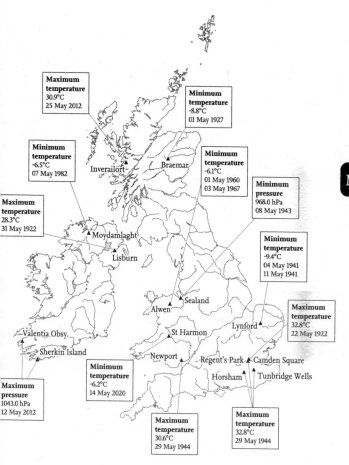

Maximum
temperature
30.9°C
25 May 2012

Minimum
temperature
-8.8°C
01 May 1927

Minimum
temperature
-6.5°C
07 May 1982

Minimum
temperature
-6.1°C
01 May 1960
03 May 1967

Minimum
pressure
968.0 hPa
08 May 1943

Maximum
temperature
28.3°C
31 May 1922

Inverailort

Braemar

Moydamlaght

Lisburn

Minimum
temperature
-9.4°C
04 May 1941
11 May 1941

Maximum
temperature
32.8°C
22 May 1922

Alwen   Sealand

Lynford

Valentia Obsy.

St Harmon

Regent's Park   Camden Square

Sherkin Island

Newport

Horsham   Tunbridge Wells

Maximum
pressure
1043.0 hPa
12 May 2012

Minimum
temperature
-6.2°C
14 May 2020

Maximum
temperature
30.6°C
29 May 1944

Maximum
temperature
32.8°C
29 May 1944

M

The Weather in May 2023

Observation	Location	Date
Max. temperature 25.1°C	Porthmadog (Gwynedd)	30 May
Min. temperature -2.2°C	Loch Glascarnoch (Ross & Cromarty)	2 May
Most rainfall 43.6 mm	Harestock Sewage Works (Hampshire)	9 May
Most sunshine 16.2 hrs	Tiree (Argyll)	30 May

Supercell
A supercell is an extremely violent, persistent thunderstorm that is marked by an extremely large, rotating updraught or 'mesocyclone'. Supercells are accompanied by heavy rain, large hail and frequent cloud-to-ground lightning discharges. The updraught may extend as high as 15 km into the atmosphere. The updraught is accompanied by strong downdraughts, but the two streams of air are separated horizontally in space, and this is one reason for a supercell's long lifetime (sometimes many hours) when compared with 'ordinary' thunderstorms, which may persist for about one hour. Cool downdraught air often bleeds into the mesocyclone and is the site of the formation of tornadoes.

The month was generally settled, with the exception of scattered thunderstorms in the second week. Locally intense downpours on the 5th resulted in surface flooding in Lincolnshire, and from the 7th severe weather warnings were issued for much of southern and central England. The 9th saw severe showers and thunderstorms develop, starting in south-western England and the south coast. Harestock in Hampshire saw 43.6 mm of rain. Flooding caused problems in various villages across Devon, and in North Cadbury, Somerset, evacuated residents sheltered in the village hall.

Over the following two days the rain spread to the Midlands, Wales, London and East Anglia, with reports of floods in Stevenage and elsewhere across Hertfordshire on the 11th. Some rain continued over the next few days and temperatures remained low, with frosts reported in all regions as late as the 16th.

The rest of the month was drier, driven by a weak jet stream and an extended area of high pressure across the UK for much of the period. Barring a band of rain that swept across Scotland and Northern Ireland on the 20th, very little rain fell at all from the 15th onwards. As the ground dried out, wildfires were reported across Wales on the 21st and 22nd, and in the last week of the month there were further fires in places from Dartmoor in the south-west to Marsden Moor in West Yorkshire. By the end of the month, maximum temperatures had breached 25.1°C at Porthmadog in Gwynedd, and all parts of the UK received in excess of 25 daily hours of sunshine.

Overall, the settled, anticyclonic conditions meant a drier and warmer month than average. The mean temperature for the UK was 1°C above average, at 11.6°C. Sunshine was at 108 per cent of average, with western England and Wales receiving more sun than northern and eastern areas. Although the south of England from Devon to Norfolk saw marginally above average rainfall, much of the northern and western UK was far drier than normal, and the country overall received just 55 per cent of its average rain.

Sunrise and Sunset 2024

Location	Date	Rise	Azimuth °	Set	Azimuth °
Belfast					
	1 May (Wed)	04:44	62	19:59	299
	11 May (Sat)	04:24	56	20:17	304
	21 May (Tue)	04:08	52	20:34	308
	31 May (Fri)	03:55	48	20:48	312
Cardiff					
	1 May (Wed)	04:44	64	19:37	296
	11 May (Sat)	04:27	59	19:53	301
	21 May (Tue)	04:12	55	20:07	305
	31 May (Fri)	04:02	52	20:20	308
Edinburgh					
	1 May (Wed)	04:28	61	19:53	300
	11 May (Sat)	04:07	55	20:12	305
	21 May (Tue)	03:49	50	20:31	310
	31 May (Fri)	03:36	46	20:46	314
London					
	1 May (Wed)	04:32	64	19:25	296
	11 May (Sat)	04:14	59	19:41	301
	21 May (Tue)	04:00	55	19:56	305
	31 May (Fri)	03:49	52	20:09	309

Note that all times are in Universal Time (UT), otherwise known as Greenwich Mean Time (GMT). These times do not take Summer Time (BST) into account.

Moonrise and Moonset 2024

Location	Date	Rise	Azimuth °	Set	Azimuth °
Belfast					
	1 May (Wed)	02:58	133	10:24	229
	11 May (Sat)	05:55	35	00:16	235
	21 May (Tue)	18:35	120	03:04	246
	31 May (Fri)	01:44	105	12:46	261
Cardiff					
	1 May (Wed)	02:28	130	10:30	233
	11 May (Sat)	06:12	40	–	–
				00:32	319
	21 May (Tue)	18:11	118	03:02	248
	31 May (Fri)	01:27	104	12:38	261
Edinburgh					
	1 May (Wed)	02:56	135	10:03	228
	11 May (Sat)	05:29	32	00:19	328
	21 May (Tue)	18:29	122	02:49	245
	31 May (Fri)	01:35	105	12:33	260
London					
	1 May (Wed)	02:17	130	10:16	233
	11 May (Sat)	05:58	40	–	–
				00:21	319
	21 May (Tue)	17:59	118	02:50	248
	31 May (Fri)	01:16	104	12:25	261

M

Note that all times are in Universal Time (UT), otherwise known as Greenwich Mean Time (GMT). These times do not take Summer Time (BST) into account.

Twilight Diagrams 2024

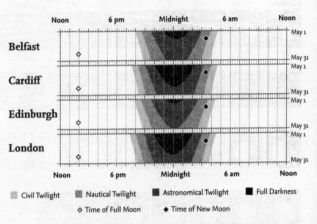

| Civil Twilight | Nautical Twilight | Astronomical Twilight | Full Darkness |
| ◇ Time of Full Moon | ◆ Time of New Moon |

The exact times of the Moon's major phases are shown on the diagrams opposite.

Adiabatic

Any process in which heat does not enter or leave the system. Air rising in the troposphere generally cools at an adiabatic rate, because it does not lose heat to its surroundings. The fall in temperature is solely because of its expansion: its increase in volume, because of the decrease in pressure with increasing altitude.

The Moon's Phases and Ages 2024

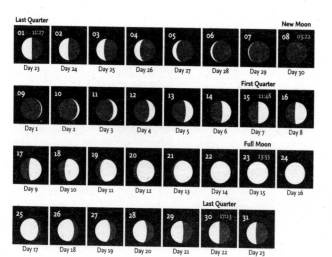

Last Quarter

01 11:27	02	03	04	05	06	07	08 03:22
Day 23	Day 24	Day 25	Day 26	Day 27	Day 28	Day 29	Day 30

New Moon

First Quarter

09	10	11	12	13	14	15 11:48	16
Day 1	Day 2	Day 3	Day 4	Day 5	Day 6	Day 7	Day 8

Full Moon

17	18	19	20	21	22	23 13:53	24
Day 9	Day 10	Day 11	Day 12	Day 13	Day 14	Day 15	Day 16

Last Quarter

25	26	27	28	29	30 17:13	31
Day 17	Day 18	Day 19	Day 20	Day 21	Day 22	Day 23

M

Hurricane
The term used for a tropical cyclone in the North Atlantic Ocean or eastern Pacific Ocean. Hurricanes (indeed all tropical cyclones) are driven by high sea-surface temperatures, and cannot occur over the British Isles.

May – In this month

2 May 1952 – The world's first jet airliner, the De Havilland Comet, made its maiden flight. It travelled from London to Johannesburg in 23 hours, changing transportation for ever.

3 May 1715 – The last total solar eclipse visible in London for almost 900 years occurred. It was known as Halley's Eclipse after the astronomer Edmond Halley, who predicted it to within four minutes of accuracy. Totality was observable from Cornwall to Norfolk and Lincolnshire, as well is in parts of Ireland.

6 May 1844 – The Glaciarium, the world's first permanent artificial ice rink, opened in London. Rather than water ice, it used a mixture of pig fat and other chemicals, and by the end of the year the smell had caused it to close. The first artificial water ice rink wouldn't open for another thirty years, in January 1874.

7 May 2000 – The Bracknell Storm caused torrential rain over a small area. This highly localised storm included hailstones up to 1.5 cm in diameter, and 65 mm of rain in less than an hour. Some flooding and other damage occurred, but neighbouring towns were unaffected.

9 May 1784 – The HMS Crocodile sank off the Devon coast, carrying 170 men. The ship was returning from India when, after entering the English Channel, it encountered thick fog and struck the rocks at Prawle Point. All of the crew was saved.

21 May 2021 – High winds across the UK resulted in a new record for the share of national electricity generation supplied by wind power, at 62.5 per cent.

22 May 1870 – Observers across Britain and Ireland reported an unusual colouration of the sun, ranging from 'dark red' to 'pink inclining to purple'. The sky was described as 'hazy' and some even described seeing sunspots on the solar disc. This was caused by a major wildfire in Canada five days previously, which sent a huge plume of smoke particles into the upper atmosphere, where they were then carried eastwards over Europe.

M

22 May 2009 – Whitelee Wind Farm opened in Scotland. With 215 turbines, it was the largest on-shore wind farm in Europe.

27 May 1774 – Francis Beaufort, inventor of the Beaufort Scale to measure wind speed, was born.

Fen Blow

The flat, marshy lands inland of the Wash, occupying parts of Cambridgeshire, Norfolk and Lincolnshire, were once known collectively as 'The Fens'. They have been extensively farmed for generations, especially since some have been artificially drained, initially by Dutch engineers. Tilling of the soil has resulted in a fine, rich soil, ideal for many crops. Unfortunately, this has also left the soil at the mercy of the wind. Strong winds are able to lift the topsoil into the air, producing choking dust storms. The soil may be carried very long distances, in a 'Fen Blow'. Although not as extreme, this is reminiscent of the situation that prevailed in the American 'Dust Bowl' in the 1930s.

Fen Blows tend to occur in spring, between March and mid-May, when the soil has been cultivated and is thus in a fine state and easily carried off by the wind. Naturally, a long dry spell of weather makes it even more likely to happen. Serious Fen Blows have occurred in 1955, 1968, 1972, 1978, 2002 and 2004. For farmers, it is disastrous when the soil has been ploughed and planted with seed, perhaps with the application of fertiliser; the wind not only lifts the topsoil, but also carries off the seed and fertiliser. Young crops may also be subjected to a form of 'sand-blasting' by the particles of soil.

Any wind over 25 mph (340 kph/22 knots) is capable of lifting off the topsoil, and as much as 5 cm has sometimes

A photograph showing the Fen Blow in 1955, caused by ploughing damage to the topsoil.

This is a classic image from the time of the great American Dust Bowl in April 1936. It shows a farmer in Cimarron County, Oklahoma, and his two sons, struggling against the dust.

M

been removed. If all of this material comes to rest in drainage ditches they quickly become clogged, worsening the situation for farmers further. The situation has been exacerbated by the fact that many hedges have been removed, so there is nothing to act as any form of windbreak. Once the seeds have germinated and put forth roots, these tend to stabilise the soil and make it less able to move, although tender young plants are also at risk of being blown away.

Probably the worst Fen Blow occurred on 4 May 1955. There had been a dry spell and then a high wind sprang up, with gusts reaching as much as 65 mph (105 kph/56.5 knots). The soil became a choking, brown cloud.

The Dust Bowl in America and Canada was similarly man-made. It arose through a failure to understand the dry-land ecology of the High Plains; farming removed the deep-rooted grasses, which had held moisture in the soil during dry spells. When drought hit in the mid-1930s, the degraded soil couldn't retain any moisture and blew away, ruining several harvests over consecutive years. Conditions were worst in northern Texas and western Oklahoma, where tens of thousands of poverty-stricken farmers had to abandon their farms.

June

Introduction

Although meteorologists regard the month of June as marking the start of the summer season, and it may see the longest day at the summer solstice (20 June in 2024), in Britain it includes the hottest day just about one year in four. This is about the same as the frequency in the month of August. The hottest day is most common in July. To the general public, June may have come to be associated with the catchphrase 'flaming June', but it is not often accompanied by particularly hot weather, and is only very rarely warmer than July. It shows little sign of getting warmer over time. During the last three centuries, the month has been about as warm as September.

In the middle of the month, the weather (when regarded as showing five seasons), tends to change quite suddenly from 'spring and early summer' to 'high summer'. This sees a quite sudden resumption of predominant westerly conditions from the mixed regimes that prevailed earlier in the year. However, the reduced temperature contrast at this time of year over the Atlantic results in weaker winds and the depressions that arrive from the Atlantic are slow-moving, so any accompanying rain tends to linger and be slow to move away. This striking change in the prevailing weather is often considered to be the start of the 'European monsoon', marked by slow-moving depressions that cross the British Isles and track towards the Baltic or farther northward towards Scandinavia.

High summer – 18 June to 9 September

There is a significant change in the overall circulation in early June, from the changeable situation that has prevailed over the preceding three months, when northerly and easterly airflow often predominated. In mid-June, the dominant westerly circulation is restored, with persistent westerly and north-westerly winds and their accompanying depressions. More settled conditions sometimes arise, when the Azores High extends a ridge of high pressure towards western Europe and occasionally over the British Isles, resulting in a fine, warm, dry summer.

Previous page: Flaming June, *the most famous painting (often regarded as his masterpiece) by Frederic Leighton, painted in 1895.*

Weather Extremes in June

Country	Temp.	Location	Date
Maximum temperature			
England	35.6°C	Camden Square (London)	29 Jun 1957
		Mayflower Park (Southampton)	28 Jun 1976
Wales	33.7°C	Machynlleth (Powys)	18 Jun 2000
Scotland	32.2°C	Ochtertyre (Perth & Kinross)	18 Jun 1893
Northern Ireland	30.8°C	Knockareven (Co. Fermanagh)	30 Jun 1976
Minimum temperature			
England	-5.6°C	Santon Downham (Norfolk)	1 Jun 1962 3 Jun 1962
Wales	-4.0°C	St Harmon (Powys)	8 Jun 1985
Scotland	-5.6°C	Dalwhinnie (Highland)	9 Jun 1955
Northern Ireland	-2.4°C	Lough Navar Forest (Co. Fermanagh)	4 Jun 1991

Country	Pressure	Location	Date
Maximum pressure			
Republic of Ireland	1043.1 hPa	Clones (Co. Monaghan)	14 Jun 1959
Minimum pressure			
Scotland	968.4 hPa	Lerwick (Shetland)	28 Jun 1938

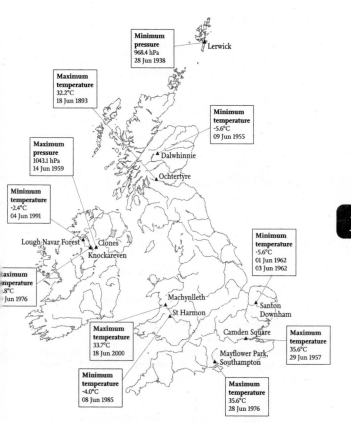

Minimum
pressure
968.4 hPa
28 Jun 1938

Lerwick

Maximum
temperature
32.2°C
18 Jun 1893

Minimum
temperature
-5.6°C
09 Jun 1955

Maximum
pressure
1043.1 hPa
14 Jun 1959

▲ Dalwhinnie

Ochtertyre

Minimum
temperature
-2.4°C
04 Jun 1991

Lough Navar Forest ▲ Clones
Knockareven

Minimum
temperature
-5.6°C
01 Jun 1962
03 Jun 1962

aximum
mperature
8°C
Jun 1976

Machynlleth
St Harmon

Santon
Downham

Maximum
temperature
33.7°C
18 Jun 2000

Camden Square

Maximum
temperature
35.6°C
29 Jun 1957

Mayflower Park,
Southampton

Minimum
temperature
-4.0°C
08 Jun 1985

Maximum
temperature
35.6°C
28 Jun 1976

J

The Weather in June 2023

Observation	Location	Date
Max. temperature 32.2°C	Chertsey, Abbey Mead (Surrey)	10 June
	Coningsby (Lincolnshire)	25 June
Min. temperature -2.6°C	Kinbrace (Sutherland)	2 June
Most rainfall 70.4 mm	Wiley Sike No. 2 (Cumbria)	18 June
Most sunshine 16.9 hrs	Loch of Hundland (Orkney)	15 June
Highest gust 54 mph (47 Kt)	South Uist Range (Western Isles)	24 June

The month began uneventfully, with the settled anticyclonic conditions continuing. As the high pressure moved away, warm continental air affected all parts, with mean, maximum and minimum temperature records broken in all regions. Highs exceeded 32°C in Surrey on the 10th.

More humid air brought rain and thunderstorms in the West Midlands and north-west England. Rain spread over the next two days, and on the 13th flooding was reported from London to as far north as Inverness. Flash floods closed the M6 for a short period, and the West Highland Line between Fort William and Crianlarich was closed for several days. In addition, landslides closed several roads in Scotland. Many places in the east of England had seen almost no rainfall by this point in the month, however.

The following few days were more settled but still very warm before more convective systems developing from the Atlantic brought thundery downpours across the whole of the UK between the 16th and the 22nd. Houses were flooded in Radcliffe, Greater Manchester, and roads and properties impacted from Norfolk to Devon. Wiley Sike (the bombing range at Spadeadam in Cumbria) received 70.4 mm of rain on the 18th. There were delays on the West Coast Main Line between London and Liverpool on the 20th, and the 22nd brought disruption to rail services in the Glasgow area.

The rest of the month remained changeable but humidity and high temperatures dominated, with Coningsby in Lincolnshire recording a maximum of 32.2°C on the 25th. More rain brought some respite from the heat, with temperatures everywhere much cooler from the 26th to the end of the month, but humidity remained high.

Overall, this was the warmest June since 1884 and the sunniest since 1957, with the UK mean temperature for the month 2.5°C higher than average. Unprecedented warm, humid weather saw maxima in excess of 25°C for over a fortnight. Average temperatures were exceeded in all areas, with parts of western Scotland seeing mean maximum temperatures reach 4°C above average. It was almost the warmest June globally on record, beating the record from 2019. Although the north-west and parts of the Midlands received slightly above average rainfall, Wales, the south and the east were very dry, and as a whole the UK received only 68 per cent of its average rainfall.

J

Nacreous clouds
Brilliantly coloured clouds (also known as 'mother-of-pearl' clouds) that are occasionally seen at sunset or sunrise. They occur in the lowest region of the stratosphere at altitudes of 15–30 kilometres. They arise when wave motion at altitude causes water vapour to freeze onto suitable nuclei at very low temperatures (below -83°C).

Sunrise and Sunset 2024

Location	Date	Rise	Azimuth °	Set	Azimuth °
Belfast					
	1 Jun (Sat)	03:55	48	20:50	312
	11 Jun (Tue)	03:48	46	21:00	314
	21 Jun (Fri)	03:47	45	21:04	315
	30 Jun (Sun)	03:52	46	21:03	314
Cardiff					
	1 Jun (Sat)	04:01	51	20:21	309
	11 Jun (Tue)	03:56	50	20:30	311
	21 Jun (Fri)	03:56	49	20:34	311
	30 Jun (Sun)	04:00	50	20:33	310
Edinburgh					
	1 Jun (Sat)	03:35	46	20:47	314
	11 Jun (Tue)	03:28	44	20:58	316
	21 Jun (Fri)	03:27	43	21:03	317
	30 Jun (Sun)	03:31	44	21:01	316
London					
	1 Jun (Sat)	03:49	51	20:10	309
	11 Jun (Tue)	03:43	49	20:18	311
	21 Jun (Fri)	03:43	49	20:23	311
	30 Jun (Sun)	03:47	49	20:22	311

Note that all times are in Universal Time (UT), otherwise known as Greenwich Mean Time (GMT). These times do not take Summer Time (BST) into account.

Moonrise and Moonset 2024

Location	Date	Rise	Azimuth °	Set	Azimuth °
Belfast					
	1 Jun (Sat)	01:52	93	14:16	27
	11 Jun (Tue)	08:41	56	00:25	307
	21 Jun (Fri)	21:36	145	02:23	217
	30 Jun (Sun)	00:18	73	15:02	294
Cardiff					
	1 Jun (Sat)	01:40	93	14:03	273
	11 Jun (Tue)	08:44	59	–	–
				00:15	296
	21 Jun (Fri)	20:57	140	02:37	222
	30 Jun (Sun)	00:13	74	14:41	292
Edinburgh					
	1 Jun (Sat)	01:42	93	14:05	273
	11 Jun (Tue)	08:23	55	00:22	308
	21 Jun (Fri)	21:40	148	01:58	214
	30 Jun (Sun)	00:04	72	14:55	295
London					
	1 Jun (Sat)	01:28	93	13:51	273
	11 Jun (Tue)	08:31	59	–	–
				00:04	296
	21 Jun (Fri)	20:45	140	02:22	222
	30 Jun (Sun)	00:01	74	14:29	292

Note that all times are in Universal Time (UT), otherwise known as Greenwich Mean Time (GMT). These times do not take Summer Time (BST) into account.

Twilight Diagrams 2024

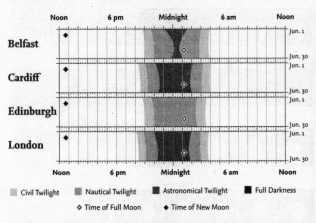

The exact times of the Moon's major phases are shown on the diagrams opposite.

Jet streams

Jet streams are narrow ribbons of fast-moving air, typically hundreds of kilometres wide and a few kilometres in depth. The most important one for British weather is the Polar Front Jet Stream, a westerly wind that flows right round the Earth. It is driven by the great temperature difference between the cold polar air and warmer air closer to the equator. Fluctuations in latitude are primarily caused by the flow across the Rockies in North America. These fluctuations in latitude, known as Rossby waves, spread right across the continental United States and across the Atlantic – and even farther. The jet stream has a great effect on the strength of depressions and also on their paths. It may cause depressions to sometimes pass directly across the British Isles and sometimes to the north or south of them.

The Moon's Phases and Ages 2024

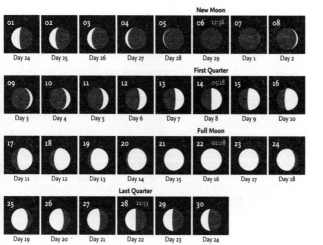

New Moon

01	02	03	04	05	06 12:38	07	08
Day 24	Day 25	Day 26	Day 27	Day 28	Day 29	Day 1	Day 2

First Quarter

09	10	11	12	13	14 05:18	15	16
Day 3	Day 4	Day 5	Day 6	Day 7	Day 8	Day 9	Day 10

Full Moon

17	18	19	20	21	22 01:08	23	24
Day 11	Day 12	Day 13	Day 14	Day 15	Day 16	Day 17	Day 18

Last Quarter

25	26	27	28 21:53	29	30
Day 19	Day 20	Day 21	Day 22	Day 23	Day 24

J

Gulf Stream

A warm-water current on the western side of the North Atlantic
Ocean. It extends along the eastern seaboard of the United States
from the Gulf of Mexico to Cape Hatteras. It then turns eastwards
and becomes the North Atlantic Current. The warm water affecting
the British Isles is a branch of this current, known as the North
Atlantic Drift (often incorrectly called the Gulf Stream). This branch
leaves the main current in the mid-Atlantic and, passing west of
Ireland, heads up towards Norway and the Arctic Ocean.

June – In this month

1 June 1831 – James Clark Ross discovered the magnetic North Pole.

1 June 1910 – Robert Falcon Scott's expedition to the South Pole left England. Scott would be beaten to the pole by Roald Amundsen on 14 December 1911, arriving five weeks later. Scott and the other members of the expedition all died on the way back across Antarctica.

2 June 1975 – The most significant June snowfall in living memory fell in many parts of the country. A cricket match between Essex and Kent at Colchester was interrupted, while the match between Derbyshire and Lancashire at Buxton was called off after 2.5 cm of snow settled on the outfield.

8 June 1924 – George Mallory and Andrew Irvine were last seen 'going strong for the top' of Mount Everest by fellow climber Noel Odell. The two mountaineers were never seen alive again.

10 June 2022 – The discovery of the wreck of HMS Gloucester off the Norfolk coast was announced. The ship was wrecked in 1682 while carrying James Stuart, Duke of York (the future James II & VII of England and Scotland). Although the wreck was discovered in 2007, the announcement was delayed for fifteen years for security reasons.

18 June 1914 – A thunderstorm near Carrbridge in the Scottish Highlands led to flooding and the destruction of a railway bridge. A resultant train accident killed five people.

23 June 1976 – The 1976 summer heatwave began. From 23 June there followed fifteen consecutive days when temperatures reached 32.3°C somewhere in England. Combined with a severe drought and high temperatures over the next two months, the summer heatwave is estimated to have caused 20% excess deaths.

29 June 2022 – The Climate Change Committee published its largest UK climate progress report to date. In it, the committee warned that the UK will fail to achieve its target of net zero by 2050 under current policies.

30 June 1908 – The Tunguska Event, a 12-megaton explosion, occurred near the Podkamennaya Tunguska River in Russia, flattening an estimated 80 million trees over a 2,150 square-kilometre area of forest. The event is generally attributed to the explosion of an asteroid about 50–60 m in size, at an altitude of 5–10 km. In Britain, this caused spectacular sunsets and bright nights.

D-Day

2024 sees the 80th anniversary of the D-Day landings (operation Overlord) and, of course, the weather was all-important for the success of the landings. The operation had to be carried out at low tide so that beach defences would be visible to the landing craft, and there had to be minimal cloud cover, so that aircraft operations and the parachute drops could occur in relative safety. On the other hand, the landing needed to be at dawn, with some moonlight but not a full moon, to provide adequate cover of darkness while the fleet of ships crossed the Channel. The requirements for moon and tide restricted the possible dates in Spring/Summer 1944 to the beginning of May, and the first and third weeks of June. Logistical requirements wrote May off, which left the forecasters a very small range of options from which to choose the most favourable.

Unfortunately, the British and American armed forces each had its own forecasting unit. There was considerable disagreement over the weather to be expected over the early days of June. The American forecasters tended to follow the 'analogue' method, whereby one searched for a previous occasion, or occasions, when conditions were similar, and predicted forthcoming weather from what had happened previously. The British contingent, by contrast, tended to rely upon the plotted conditions as reported by weather stations and extrapolated the weather from that.

For early June, the American forecasts suggested that there would be minimal cloud cover and winds for 4 and 5 June. The general British view was that cloud cover would be considerable on those days, but that a short improvement would occur on the following day, 6 June, with little cloud cover and favourable winds. Any delay beyond this would mean putting off the operation for a fortnight to satisfy the tidal conditions. It was up to the leading British forecaster, Group Captain James Stagg, to brief General Eisenhower, who was in charge of the whole operation.

Eisenhower went by the British advice, and delayed the start of operations by one day, from 5 to 6 June. This was an agonising decision; hundreds of ships and thousands of soldiers were already at sea and had to turn back. Moreover, there was no guarantee that the weather would be any better the following day. However, Eisenhower's call proved to be crucial. The weather was indeed better on the second day, allowing the operation to take place. The conditions were not perfect, especially in the eastern (British) sector, where the wind and waves were somewhat stronger than desirable. Nevertheless, the surf proved to be passable and, although not ideal, the landing operation was a general success.

Shortly afterwards came a demonstration of what could have happened had the weather not held. Of the two artificial 'Mulberry' harbours, established off the beaches at Gold Beach (in the British sector) and Omaha Beach (in the American sector) the western one (Mulberry B, off Omaha Beach) was irreparably damaged by a violent storm from the north-east on June 19, and had to be abandoned. This meant the entire invasion force was dependent on the other harbour (Mulberry A) to transport heavy equipment. Luckily this remained operational, providing deep water for the large vessels to unload, and was used for at least 10 months after D-Day, with more than 2.5 million men and 4 million tonnes of materiel landed successfully.

Had the decision been taken to attempt the amphibious assault a day earlier, or a fortnight later, countless lives and equipment might have been lost before even reaching the French shore. As it was, wind, moon and tide were all favourable, granting the allied forces safe passage.

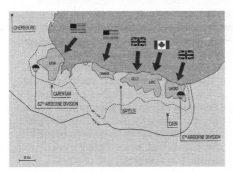

A map of the D-Day landings, showing the named beaches. The shaded areas are those controlled by the Allied forces at midnight on 6 June 1944.

J

Noctilucent Clouds

For about one month on either side of midsummer, it is sometimes possible at night (even in the middle of the night) to see glowing clouds when looking in the direction of the pole. These are very high clouds, known as noctilucent clouds. (The name means 'night-shining'.) From Britain they are seen in the general direction of the North Pole and tend to be mostly observed from Scotland, although on occasions there are major displays that may be seen from anywhere in the country.

The period of NLC visibility is actually about one month before the summer solstice until roughly two months after the solstice. It is believed that this inequality is caused by the time for the atmosphere to cool sufficiently for the necessary temperatures to be reached at that altitude. The 'spike' in activity in 2022 may be related to water vapour introduced by the SpaceX launch of the Globalstar satellite on June 19.

NLC are seen from both hemispheres, although less frequently reported from the south, probably because of the low population density and the small amount of land from which they may be observed.

These clouds (NLC) are seen when the observer is in darkness, but the clouds – which are the highest in the atmosphere at about 185 kilometres, far above all other clouds – are still illuminated by the Sun, which is below the northern horizon. (The 'normal' clouds, such as those described on pages 233–244 are found in the lowest layer of the atmosphere, the troposphere, which, even in the tropics, extends to an altitude of no more than 20 km.)

Noctilucent clouds consist of ice crystals, believed to form around particles of meteoritic dust arriving from space. Although they are known to be formed of ice, the origin of the water that freezes into ice crystals is still unknown and the subject of great debate. For many years it was believed that the water also came 'from outside' and was brought by comets and other bodies. There is another possibility. Although water vapour cannot be carried to such great heights, it is possible that water vapour may be created by the breakdown of methane gas, which can rise freely to such extreme altitudes and then be broken down by radiation from the Sun.

Noctilucent clouds, photographed by Alan Tough from Nairn in Scotland, on 31 May 2020, at 00:28.

J

July

Introduction

The saying that 'the English summer consists of three fine days and a thunderstorm' has been ascribed to both the kings Charles II and George III. However, it is probably a proverbial piece of weather lore, the origins of which are lost. A succession of hot days and humid air certainly provides the conditions for the formation of giant cumulonimbus clouds and thunderstorms with the accompanying torrential rain or even hail. Such a situation usually ends a heatwave (at least for a day or so) and is very typical of British weather.

July is definitely associated with high summer, and it is usually the hottest month of the year. It frequently includes the hottest day of the year. This occurs about 44 per cent of the time, the remaining percentage of hottest days being more-or-less equally divided between June and August. Most recently, in 2022, England, Wales and Scotland all recorded new record temperatures in this month, with Coningsby in Lincolnshire reaching 40.3°C.

Longer ago, the year 1976 was the 'drought year', when, beginning in late June, there were ten weeks of sunshine and practically no rain. (It was actually the second driest summer of the twentieth century, after 1995.) A very few locations were fortunate and had some rain, although this was generally less than half of the usual rainfall for the month of July. A high-pressure 'block' over the UK diverted the normal succession of depressions from the Atlantic, and their accompanying rain, south towards the Mediterranean. The drought only came to an end (in late August) just after the government appointed a Drought Minister.

The prolonged drought was actually the result of a very long period of reduced rainfall. In the preceding year, 1975, both the summer and autumn were dry, the winter 1975–76 was particularly dry and then so was the spring of 1976. (Here we use the three-month 'meteorological' seasons, rather than the five seasons that we discuss on page 11.) Certain areas of the country actually experienced months with no rain. Portions of south-west England had no rain at all in July and in the first half of August 1976.

July 2006 was the warmest calendar month ever recorded in the Central England Temperature series, which began in 1659. There was an unusual high pressure region over northern Europe and a persistent airstream from the south affecting the British Isles.

J

Icelandic Low
A semi-permanent feature of the distribution of pressure over the North Atlantic. Unlike the more-or-less permanent Azores High, it largely arises because depressions (low-pressure systems) frequently pass across the area. A similar low-pressure area exists over the northern Pacific Ocean, often known as the 'Aleutian Low'.

Weather Extremes in July

Country	Temp.	Location	Date
Maximum temperature			
England	40.3°C	Coningsby (Lincolnshire)	19 Jul 2022
Wales	37.1°C	Hawarden Airport (Flintshire)	18 Jul 2022
Scotland	34.8°C	Charterhall (Scottish Borders)	19 Jul 2022
Northern Ireland	31.3°C	Castlederg (Co. Tyrone)	21 Jul 2021
Minimum temperature			
England	-1.7°C	Kielder Castle (Northumberland)	17 Jul 1965
Wales	-1.5°C	St Harmon (Powys)	3 Jul 1984
Scotland	-2.5°C	Lagganlia (Inverness-shire)	15 Jul 1977
Northern Ireland	-1.1°C	Lislap Forest (Co. Tyrone)	17 Jul 1971

Country	Pressure	Location	Date
Maximum pressure			
Scotland	1039.2 hPa	Aboyne (Aberdeenshire)	16 Jul 1996
Minimum pressure			
Scotland	967.9 hPa	Sule Skerry (Northern Isles)	8 Jul 1964

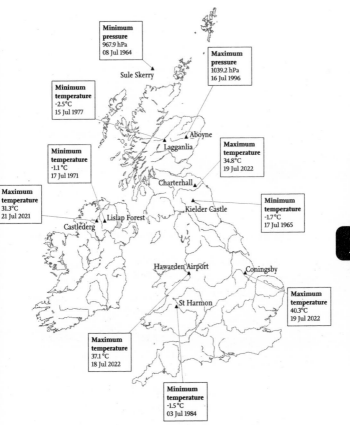

Minimum pressure
967.9 hPa
08 Jul 1964

Maximum pressure
1039.2 hPa
16 Jul 1996

Minimum temperature
-2.5°C
15 Jul 1977

Sule Skerry

▲ Aboyne

Lagganlia

Maximum temperature
34.8°C
19 Jul 2022

Minimum temperature
-1.1°C
17 Jul 1971

Charterhall ▲

Minimum temperature
-1.7°C
17 Jul 1965

Maximum temperature
31.3°C
21 Jul 2021

Lislap Forest ▲

Kielder Castle ▲

Castlederg

Hawarden Airport ▲

Coningsby ▲

St Harmon ▲

Maximum temperature
40.3°C
19 Jul 2022

Maximum temperature
37.1°C
18 Jul 2022

Minimum temperature
-1.5°C
03 Jul 1984

J

The Weather in July 2023

Observation	Location	Date
Max. temperature 30.2°C	Chertsey, Abbey Mead (Surrey)	7 July
Min. temperature 1.2°C	Loch Glascarnoch (Ross & Cromarty)	26 July
Most rainfall 110.9 mm	White Barrow (Devon)	22 July
Most sunshine 16.1 hrs	Lerwick (Shetland)	8 July
Highest gust 79 mph (69 Kt)	Needles Old Battery (Isle of Wight)	15 July

July was wetter and cooler than normal, with several summer storms. The jet stream was located further to the south and there was an abnormally large pressure gradient from the Azores High to the low over Scandinavia, resulting in autumnal weather. This being festival season, several events were adversely affected. On the 6th, the Tiree music festival was cancelled, and localised flooding made various other events washouts.

That weekend was the hottest of the month, with a high of 30.2°C recorded at Chertsey in Surrey. However, such high temperatures quickly broke down into thunderstorms. On the 8th, roads were flooded in Liverpool and north-east Wales, and there was surface flooding in Derbyshire and the West Midlands. Storms on the 9th were followed by more heavy rain over the next two days, resulting in disruption to rail services in central Scotland on the 10th and in Birmingham on the 11th.

Further disruption followed over the next two weeks as a series of fronts travelled west-to-east across the country. A particularly deep depression brought thunderstorms and high winds on the 15th; the south coast saw gusts touching 80 mph, while lightning caused a large-scale electricity outage in Northern Ireland on the 15th. Both Northern Ireland and Newcastle saw roads flooded and trees downed.

The following weekend saw another series of depressions bring the most persistent rain. A total of 110.9 mm fell at White Barrow in Devon on the 22nd and the Garstang flood basin was opened to ease the pressure on local rivers. By the end of the month many drainage systems were at capacity, and even slightly unsettled weather saw localised flooding in Scarborough.

Overall, July was much more unsettled than June, with lower temperatures, higher winds and more rainfall. Daily maxima frequently failed to cross 20°C, and towards the end of the month minima fell, with 1.2°C recorded at Loch Glascarnoch. Lancashire, Merseyside, Manchester and parts of Northern Ireland recorded 200 per cent of their average rain. Overall, UK rainfall was 170 per cent of the average, making this the wettest July since 2009, while for Northern Ireland it was the wettest in 188 years. Sunshine was well below normal, with places in the south and west recording only 81 per cent of the normal.

Doldrums
The doldrums are a zone of reduced winds, generally located over the equatorial region, although moving north and south with the seasons. Air in the Doldrums is largely rising, because solar heating and horizontal motion across the surface is reduced or non-existent.

Sunrise and Sunset 2024

Location	Date	Rise	Azimuth °	Set	Azimuth °
Belfast					
	1 Jul (Mon)	03:52	46	21:03	314
	11 Jul (Thu)	04:03	48	20:55	312
	21 Jul (Sun)	04:17	52	20:42	308
	31 Jul (Wed)	04:33	56	20:26	304
Cardiff					
	1 Jul (Mon)	04:00	50	20:33	310
	11 Jul (Thu)	04:09	52	20:27	308
	21 Jul (Sun)	04:21	55	20:16	305
	31 Jul (Wed)	04:36	59	20:02	301
Edinburgh					
	1 Jul (Mon)	03:32	44	21:01	316
	11 Jul (Thu)	03:43	46	20:53	314
	21 Jul (Sun)	03:58	50	20:39	310
	31 Jul (Wed)	04:16	55	20:21	305
London					
	1 Jul (Mon)	03:48	49	20:22	310
	11 Jul (Thu)	03:57	51	20:16	308
	21 Jul (Sun)	04:09	55	20:05	305
	31 Jul (Wed)	04:23	59	19:50	301

Note that all times are in Universal Time (UT), otherwise known as Greenwich Mean Time (GMT). These times do not take Summer Time (BST) into account.

Moonrise and Moonset 2024

Location	Date	Rise	Azimuth °	Set	Azimuth °
Belfast					
	1 Jul (Mon)	00:28	62	16:35	305
	11 Jul (Thu)	10:13	82	23:06	272
	21 Jul (Sun)	21:31	133	03:24	221
	31 Jul (Wed)	– 23:35	– 38	18:38	325
Cardiff					
	1 Jul (Mon)	00:28	64	16:08	302
	11 Jul (Thu)	10:04	83	22:55	272
	21 Jul (Sun)	21:01	130	03:36	225
	31 Jul (Wed)	– 23:48	– 43	17:59	319
Edinburgh					
	1 Jul (Mon)	00:12	61	16:30	306
	11 Jul (Thu)	10:00	82	22:56	273
	21 Jul (Sun)	21:28	135	03:01	218
	31 Jul (Wed)	23:53	302	18:41	327
London					
	1 Jul (Mon)	00:15	64	15:57	302
	11 Jul (Thu)	09:52	83	22:43	272
	21 Jul (Sun)	20:50	130	03:23	225
	31 Jul (Wed)	– 23:35	– 43	17:48	319

Note that all times are in Universal Time (UT), otherwise known as Greenwich Mean Time (GMT). These times do not take Summer Time (BST) into account.

J

Twilight Diagrams 2024

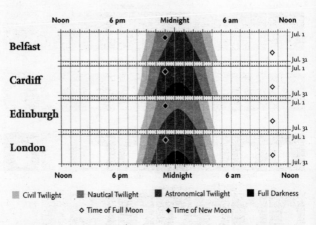

The exact times of the Moon's major phases are shown on the diagrams opposite.

Troposphere
The lowest layer in the atmosphere, in which essentially all weather occurs. It is defined by the way in which temperature declines with height, and is bounded at the top by the tropopause (an inversion at which temperatures either stabilise or begin to increase with height in the overlying stratosphere. The height of the tropopause (and thus the depth of the troposphere) increases from about 7 kilometres at the poles to 14–18 kilometres at the equator.

The Moon's Phases and Ages 2024

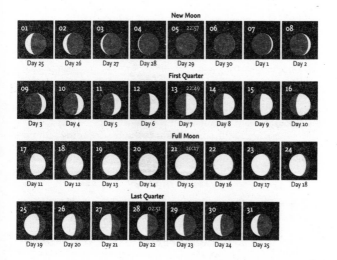

New Moon

01	02	03	04	05 22:57	06	07	08
Day 25	Day 26	Day 27	Day 28	Day 29	Day 30	Day 1	Day 2

First Quarter

09	10	11	12	13 22:49	14	15	16
Day 3	Day 4	Day 5	Day 6	Day 7	Day 8	Day 9	Day 10

Full Moon

17	18	19	20	21 10:17	22	23	24
Day 11	Day 12	Day 13	Day 14	Day 15	Day 16	Day 17	Day 18

Last Quarter

25	26	27	28 02:51	29	30	31
Day 19	Day 20	Day 21	Day 22	Day 23	Day 24	Day 25

J

Tropopause
The tropopause is the boundary between the lowest layer in
the atmosphere (the troposphere) and the next highest (the
stratosphere). It is an inversion at which temperatures either
stabilise or begin to increase with height in the overlying
stratosphere. The height of the tropopause (and thus the depth of
the troposphere) increases from about 7 kilometres at the poles to
14–18 kilometres at the equator. There are breaks in the level of the
tropopause (particularly near the location of jet streams) and these
do allow some exchange of air between the layers.

July – In this month

1 July 1714 – Parliament passed the Longitude Act, offering a prize to anyone who can solve the problem of accurately determining a ship's longitude. Prize money was subsequently paid to various inventors of lunar distance tables, watches and chronometers.

6 July 1936 – A major breach of the Manchester, Bolton and Bury Canal sent millions of gallons of water cascading into the River Irwell, 200 ft below.

9 July 1923 – A thunderstorm in south-west London contained some of the most vivid and prolonged lightning observed in this country. Nearly 7000 lightning flashes were recorded at Chelsea in six hours starting at 11pm, which equates to almost 18 flashes a minute. Many houses were struck by lightning, and some caught fire; a house at Walton-on-the-Hill was destroyed.

18 July 2022 – European bison were reintroduced to the UK for the first time in thousands of years. Three individuals were released to roam in the Kent countryside, in the hope that they would improve the local ecology.

19 July 2022 – A temperature above 40°C was recorded for the first time in the UK. A provisional Met Office reading of 40.3°C at Coningsby, Lincolnshire beat the previous record high of 38.7°C set in Cambridge in July 2019.

20 July 1929 – Two people were drowned by a meteotsunami in West Sussex. This event, a small tsunami caused by atmospheric conditions such as a rapid change of air pressure, hit the south coast between Goring and Shoreham, capsizing boats and flooding beaches.

25 July 1609 – While en route to Virginia, the English ship Sea Venture was deliberately driven ashore at Bermuda during a storm to prevent its sinking. The survivors went on to found a new colony where they landed.

28 July 2005 – An F2 tornado hit Birmingham in the early afternoon. Nineteen people were hurt, some seriously.

28 July 2021 – Orbital O2, the world's most powerful tidal turbine device, began generating electricity off the coast of Eday in the Orkney Islands.

J

The Longitude Act

In July 1714, Parliament passed the Longitude Act. The act aimed to encourage major advances towards solving the problem of calculating longitude at sea by offering monetary prizes to people who could demonstrate new technologies or methods. It also established a Commission for the Discovery of the Longitude Act, also known as the Board of Longitude, to administer the prizes.

The act was preceded by several seafaring accidents resulting from the inability to calculate longitude accurately. In the absence of accurate latitude and longitude measurements, seafarers relied on purely observational navigation and between 1550 and 1560, one in five ships travelling between Portugal and India was lost. As transatlantic trade in sugar, cotton and slaves increased, the incentive to make shipping safer increased.

The immediate catalyst, however, was the Scilly Naval Disaster of 1707. On 21 October of that year, four ships out of a fleet of 21, including the flagship Association, sank after hitting rocks west of the Scilly Isles. They had been relying on a combination of soundings and dead reckoning to calculate their longitude, and also made inaccurate readings of latitude, partly because they were prevented from taking proper sightings of the Sun by the bad weather. Believing themselves to be further south and west than they were, they set a course that was too northerly and hit the rocks. Exacerbating their navigational difficulties were the bad weather and the since-confirmed Rennell's Current, which can push ships in that area to the north.

The disaster brought the issue of longitude calculation firmly into the national interest and, seven years later, Parliament made its offer of monetary incentives to those who could solve the problem. Prizes of up to £20,000 for a method that could reliably calculate longitude to within half a degree were made available, equating to over £3 million today. The Board could also administer at their discretion smaller amounts to people who made significant contributions towards a solution.

It was already known at this stage that the Earth rotates through 15 degrees of longitude per hour (discounting the marginal difference between a solar day and a sidereal day). By comparing the local time, easily calculated by observing the Sun, with the

A rather fanciful engraving of HMS Association, just before she struck the Outer Gilstone Rock off the Scilly Islands, with the loss of her whole crew of 800 sailors and Admiral Sir Cloudesley Shovell. This disaster precipitated the passing of the Longitude Act.

J

time at a fixed reference point, one could therefore obtain a longitude reading. This had been proposed in the sixteenth century by the Dutch scientist Gemma Frisius, but the clocks of the period relied on the swinging of a pendulum to regulate time. They therefore wouldn't work on a moving ship.

The breakthrough came thanks to John Harrison, a carpenter from Yorkshire whose marine chronometer was the first clock to be able to tell the time on a moving ship. His first design was based on weighted beams connected by springs. These beams would oscillate against each other and so provide a form of movement regulated by the spring constant of the coils, rather than by gravity. His chronometer underwent several refinements, and the final 'H4' iteration used a much faster oscillating balance wheel, controlled by a spiral spring, features which remain the standard in clockwork watches today. Further advancements were made in the following years by Pierre Le Roy and, by the 1820s, all Royal Navy vessels were equipped with a chronometer. No more would anomalous wind directions or other weather phenomena be such a major cause of casualties at sea.

For his final design, Harrison received an interim payment of £10,000 in 1765. However, although he submitted the design for the headline £20,000 prize, he never received it. By then the specifications of the Longitude Act had changed, and in 1773 Parliament voted to offer him a special award of £8,750 instead.

Orographic
A term used to describe rain, or conditions that arise when air is forced to rise over high ground. The increase of rainfall (orographic rain) is a common occurrence over the mountains of Wales and Scotland.

August

Introduction

Although August is always regarded as high summer, it sees the hottest day only as frequently as June. Both have fewer such days than July. August is also a quiet month, with less wind, being similar to July in that respect. However, to many counties in the east of England it is the wettest month as many thunderstorms and their associated rain travel across from the west.

The very first weather forecast appeared in *The Times* on 1 August 1861. It was prepared by Robert FitzRoy – best known to the general public as the captain of HMS Beagle, in which Charles Darwin made his momentous voyage. FitzRoy had been appointed in 1854 as 'Statist' to the newly formed Meteorological Department of the Board of Trade. Although, initially, FitzRoy merely collated observations submitted by mariners, he then instigated a system for communicating gale warnings to sailors at various ports. He went on to start 'prognosticating' forthcoming weather, and indeed introduced the term 'weather forecasting'. This resulted in his producing, and *The Times* publishing, the first ever forecast.

This revolutionary forecast was:

General weather probable during next two days:
North – Moderate westerly wind; fine
West – Moderate south-westerly; fine
South – Fresh westerly; fine

Unfortunately, British weather being so changeable and (in those days) so unpredictable, FitzRoy's forecasts soon became inaccurate and developed into a subject of derision. The situation was not helped by those meteorologists who wanted to put the subject on a scientific basis. FitzRoy's empirical methods were particularly criticised by Francis Galton who, in effect, wished to calculate everything. (Galton is infamous for proposing and promoting eugenics, and famous for producing an equation for obtaining the optimum cup of tea.) As the physics behind meteorology were so poorly understood at that time, Galton believed that true forecasts were impossible. Nevertheless, he remained fascinated by meteorology, and produced the first weather map, discovered anticyclones and proposed a theory explaining their existence.

FitzRoy's storm warning system was discontinued, but reinstated within a short period because both the general public and mariners campaigned for the reintroduction. The weather forecasts were also stopped, and only on 1 April 1875 did *The Times* publish the first weather map, prepared by Galton (see page 77), as distinct from the first weather forecast.

August was originally the sixth month (known as *sextilis*) of the ten-month Roman calendar (which omitted the winter months and began the year in March). It became the eighth month when, in about 700 BCE the months January and February were added at the beginning of the year. It originally had 29 days, but gained 2 days when Julius Caesar created the Julian calendar in 46 BCE. It was renamed in honour of the emperor Augustus in 8 BCE. It retained 31 days when Pope Gregory XIII instigated the calendar reform leading to the Gregorian calendar, introducing the new calendar in 1582.

Until 1965, the August Bank Holiday was held at the beginning (not the end) of the month. The first Monday in August is still a public holiday in Scotland, but in England, Wales and Northern Ireland it has been moved to the last Monday in August.

A

Robert FitzRoy (1805–1865), widely regarded as the founder of modern meteorology. A sea area is named after him, and the headquarters of the Met Office are on Fitzroy Road in Exeter.

Weather Extremes in August

Country	Temp.	Location	Date
Maximum temperature			
England	38.5°C	Faversham (Kent)	10 Aug 2003
Wales	35.2°C	Hawarden Bridge (Flintshire)	2 Aug 1990
Scotland	32.9°C	Greycrook (Scottish Borders)	9 Aug 2003
Northern Ireland	30.6°C	Tandragee Ballylisk (Co. Armagh)	2 Aug 1995
Minimum temperature			
England	-2.0°C	Kielder Castle (Northumberland)	14 Aug 1994
Wales	-2.8°C	Alwen (Conwy)	29 Aug 1959
Scotland	-4.5°C	Lagganlia (Inverness-shire)	21 Aug 1973
Northern Ireland	-1.9°C	Katesbridge (Co. Down)	24 Aug 2014

Country	Pressure	Location	Date
Maximum pressure			
Scotland	1038.4 hPa	Altnaharra No. 2 (Sutherland)	31 Aug 2021
Minimum pressure			
Republic of Ireland	967.3 hPa	Shannon Airport (Co. Clare)	19 Aug 2020

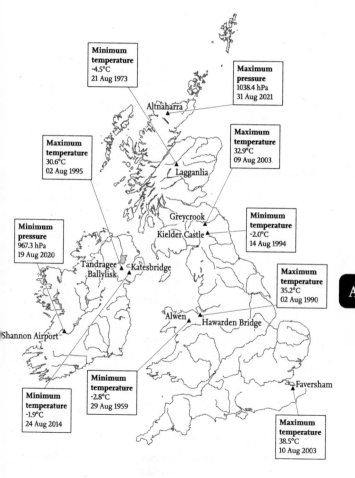

Minimum temperature
-4.5°C
21 Aug 1973

Maximum pressure
1038.4 hPa
31 Aug 2021

Altnaharra

Maximum temperature
30.6°C
02 Aug 1995

Maximum temperature
32.9°C
09 Aug 2003

Lagganlia

Greycrook

Minimum temperature
-2.0°C
14 Aug 1994

Kielder Castle

Minimum pressure
967.3 hPa
19 Aug 2020

Tandragee
Ballylisk

Katesbridge

Maximum temperature
35.2°C
02 Aug 1990

Alwen

Hawarden Bridge

Shannon Airport

Minimum temperature
-2.8°C
29 Aug 1959

Faversham

Minimum temperature
-1.9°C
24 Aug 2014

Maximum temperature
38.5°C
10 Aug 2003

A

135

The Weather in August 2022

Observation	Location	Date
Max. temperature 34.9°C	Charlwood (Surrey)	13 August
Min. temperature 0.3°C	Tulloch Bridge (Inverness-shire)	17 August
Most rainfall 146.2 mm	Holbeach (Lincolnshire)	17 August
Highest gust 75 mph (65 Kt)	South Uist Range (Western Isles)	20 August

Ionosphere
A region of the atmosphere, consisting of the upper mesosphere and part of the exosphere (from about 60–70 km to 1000 km or more) where radiation from the Sun ionises atoms and causes high electrical conductivity. The ionosphere both reflects certain radio waves back towards the surface, and blocks some wavelengths of radiation from space.

August 2022 began with Atlantic weather fronts bringing heavy rain to parts of North Wales, Cumbria and western Scotland. This led to flooding in Llanrwst on the 2nd as the Conwy river overtopped its banks. The rest of the country saw generally cloudy weather, with occasional thundery showers, and mist and fog in south-western England.

This was followed by an extended period of fine weather across much of the country. Consistent daily maxima in the low to mid 30s across England, and even into Roxburghshire in Scotland, saw an amber extreme heat warning issued for four days. On the 6th and 7th, wildfires were reported in several places around London and the south-east, followed by more wildfires in Nottinghamshire and Northamptonshire on the 12th and 13th. Only the far north of Scotland was consistently cooler during this period.

Thunderstorms brought an end to the hot spell around the middle of the month, with several places affected by floods. Fife saw disruption to roads and railways on the 14th, the roof of a supermarket in Inverness collapsed, and flash flooding was reported in Truro on the 15th, and South Wales and the East Midlands on the 16th. On the 17th the rain hit London, and several tube stations were closed due to flooding. This weekend saw the highest 24-hour rainfall anywhere, with 146.2 mm falling at Holbeach, Lincolnshire.

After a quiet spell, thunderstorms and heavy rain returned a week later across Hampshire, Greater London, Norfolk and Suffolk, but the last week of the month was drier. Patchy showers continued in parts of Northern Ireland and Scotland, but conditions were generally more settled.

In total, August was a drier than average month, with plenty of sunshine in the first half. Overall, the UK received 54 per cent of its average rain, and some places in England and Wales saw only 20 per cent of normal. The mean temperature of 16.7°C was 1.5°C above the long-term average and the fifth warmest since 1884, with a maximum of 34.9°C recorded at Charlwood in Surrey on the 13th. Sunshine totals were the fourth highest for August since 1919.

A

Sunrise and Sunset 2024

Location	Date	Rise	Azimuth °	Set	Azimuth °
Belfast					
	1 Aug (Thu)	04:35	57	20:24	303
	11 Aug (Sun)	04:53	62	20:03	298
	21 Aug (Wed)	05:11	68	19:41	292
	31 Aug (Sat)	05:30	74	19:17	286
Cardiff					
	1 Aug (Thu)	04:37	59	20:00	300
	11 Aug (Sun)	04:53	64	19:42	296
	21 Aug (Wed)	05:08	69	19:22	290
	31 Aug (Sat)	05:24	75	19:00	284
Edinburgh					
	1 Aug (Thu)	04:18	55	20:19	304
	11 Aug (Sun)	04:37	61	19:57	299
	21 Aug (Wed)	04:57	67	19:34	293
	31 Aug (Sat)	05:16	74	19:08	286
London					
	1 Aug (Thu)	04:25	59	19:49	301
	11 Aug (Sun)	04:41	64	19:31	296
	21 Aug (Wed)	04:56	69	19:10	290
	31 Aug (Sat)	05:12	75	18:49	284

Note that all times are in Universal Time (UT), otherwise known as Greenwich Mean Time (GMT). These times do not take Summer Time (BST) into account.

Moonrise and Moonset 2024

Location	Date	Rise	Azimuth °	Set	Azimuth °
Belfast					
	1 Aug (Thu)	00:19	35	19:35	324
	11 Aug (Sun)	12:55	119	21:43	238
	21 Aug (Wed)	20:25	93	07:19	261
	31 Aug (Sat)	01:46	46	18:59	308
Cardiff					
	1 Aug (Thu)	00:35	40	18:56	319
	11 Aug (Sun)	12:32	117	21:45	240
	21 Aug (Wed)	20:13	92	07:11	261
	31 Aug (Sat)	01:54	50	01:54	50
Edinburgh					
	1 Aug (Thu)	–	–	19:38	327
		23:53	32		
	11 Aug (Sun)	12:49	120	21:26	236
	21 Aug (Wed)	20:15	93	07:05	260
	31 Aug (Sat)	01:24	44	18:55	309
London					
	1 Aug (Thu)	00:22	40	18:45	319
	11 Aug (Sun)	12:20	117	21:32	240
	21 Aug (Wed)	20:01	93	06:58	261
	31 Aug (Sat)	01:41	50	18:21	305

A

Note that all times are in Universal Time (UT), otherwise known as Greenwich Mean Time (GMT). These times do not take Summer Time (BST) into account.

Twilight Diagrams 2024

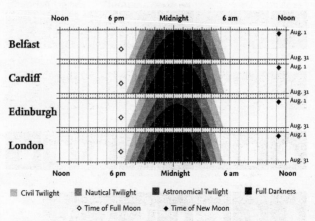

The exact times of the Moon's major phases are shown on the diagrams opposite.

Föhn effect

When humid air is forced to rise over high ground, it normally deposits some precipitation in the form of rain or snow. When the air descends on the far side of the hills, because it has lost some of its moisture, it warms at a greater rate than it cooled on its ascent. This 'Föhn effect' may cause temperatures on the leeward side of hills or mountains to be much warmer than locations at a corresponding altitude on the windward side.

The Moon's Phases and Ages 2024

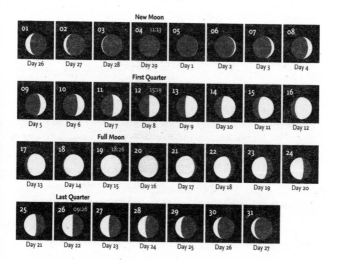

New Moon

01	02	03	04 11:13	05	06	07	08
Day 26	Day 27	Day 28	Day 29	Day 1	Day 2	Day 3	Day 4

First Quarter

09	10	11	12 15:19	13	14	15	16
Day 5	Day 6	Day 7	Day 8	Day 9	Day 10	Day 11	Day 12

Full Moon

17	18	19 18:26	20	21	22	23	24
Day 13	Day 14	Day 15	Day 16	Day 17	Day 18	Day 19	Day 20

Last Quarter

25	26 09:26	27	28	29	30	31
Day 21	Day 22	Day 23	Day 24	Day 25	Day 26	Day 27

Advection
The horizontal motion of air from one area to another. Advection of humid air over cold ground will often result in the ground cooling the overlying air, and is a common cause of mist and fog. Advection also brings sea mist onto low-lying coastal areas.

A

August – In this month

2 August 1985 – Delta Airlines flight 191 crashed after encountering a microburst while on landing approach at Dallas Fort Worth. The incident prompted fresh training and awareness about the dangers of windshear and microbursts for aircraft.

3 August 2022 – Marks and Spencer announced that they would stop selling disposable barbecues, in order to help reduce the risk of wildfires started by the objects.

8 August 1588 – The Royal Navy defeated the Spanish Armada at the Battle of Gravelines, the day after sending fireships towards the anchored Spanish fleet at Calais. After the battle, the Spanish ships were forced to flee around the north coast of Scotland, with English writers ascribing the victory to a 'Protestant wind'.

15 August 2022 – The Cabinet Office announced the launch of an emergency warning system to send alerts about severe weather events and other risks to life directly to people's mobile phones.

16 August 2016 – The world's largest ever wind farm was approved for construction off the Yorkshire coast. When complete, the array will include 300 turbines and be capable of producing 1.8 GW of electricity.

17 August 1978 – Ben Abruzzo, Maxie Anderson and Larry Newman arrived in Miserey, near Paris in Double Eagle II, completing the first ever transatlantic crossing by balloon. They had departed Presque Isle, Maine, on 11 August, and the crossing took 137 hours. Their flight was the culmination of more than a century of attempts to cross the Atlantic by balloon.

18 August 1783 – The 1783 Great Meteor was observed as it passed over the east coast of the UK before breaking up over France and Italy. Notable witnesses included Tiberius Cavallo, Alexander Aubert and Richard Lovell Edgeworth. Engravings and descriptions of the event contributed to improved understanding of meteors.

27 August 1912 – The Great Norfolk Flood saw the River Wensum rise 16.5 ft above its mean high level. Two days of heavy rainfall combined with high tides and strong north-westerly winds to break both man-made and natural riverbanks. Fifty bridges across the county were wrecked and four people died.

An August Midnight

August is known for sultry late-summer nights. With winds generally low and thunderstorms never far away, the humidity can feel oppressive. These conditions also make August the most favourable month for many insect species. This is apparent in Thomas Hardy's poem 'An August Midnight', in which the poet paints a vivid picture of himself unable to sleep, with only insects for company in the suffocating night air.

> *A shaded lamp and a waving blind,*
> *And the beat of a clock from a distant floor:*
> *On this scene enter – winged, horned, and spined –*
> *A longlegs, a moth, and a dumbledore;*
> *While 'mid my page there idly stands*
> *A sleepy fly, that rubs its hands…*
>
> *Thus meet we five, in this still place*
> *At this point of time, at this point in space.*
> *– My guests besmear my new-penned line,*
> *Or bang at the lamp and fall supine.*
> *"God's humblest, they!" I muse. Yet why?*
> *They know Earth-secrets that know not I.*

In the first stanza, Hardy sets the scene as if writing stage directions for a play: first he lays out props, movement and sound effects, and then he describes the characters as they enter the narrative. His use of familiar terms for the cranefly and bumblebee betrays a specific time and place setting, while the three insects are each given a specific characteristic – wings for the longlegs, horns for the moth and a 'spine' for the dumbledore. The fly is also identified with a physical feature, its forelimbs, but also with a particular action and attitude that anyone can recognise from a lazy summer's night. Although the punctuation and irregular rhyming pattern could imply a sense of urgent movement and direction, the metre and sibilance instead imbue the stanza with a slow, calm pace.

The second stanza continues this lyrical theme as the poet reflects on his encounter with these four creatures. The rhyming pattern is more regular, giving this verse more stability. Rhyming couplets are commonly used in the sonnet form to signal a final musing, so Hardy's use of three of them here makes clear that the focus is on reflection. The insects are described as the poet's 'guests' and equals, with their own secret knowledge and relationship with the divine – this at the same time as they are hurling themselves against his lamp and spoiling his poetry!

Hardy's Romanticism is evident throughout the poem. Both in its content and in its structure, rhyme and rhythm, it reads as a detailed observation of nature, combined with a philosophical reflection on humankind's place in relation to it.

A painting of insects by the Dutch artist Jan van Kessel (1626–1679). However, a daddy long-legs (crane-fly) and a dumbledore (bumblebee) mentioned in the poem are not shown here.

A

September

Introduction

September has always been regarded as a transitional month, between the high summer of July and August and the autumn. It is a time for late holidays and generally fairly quiet weather. The month tends to be drier than the preceding month, August, and the following one, October. Although there is often a period of windy weather, sometimes with gale-force winds, generally late in the month, the mariners' myth of 'equinoctial gales' certainly does not apply (see the discussion of this point under March, page 48).

To meteorologists, September is the first month of autumn. However, if the year is regarded as consisting of five seasons, the transition to autumn occurs a bit later, after the first week of the month, in which the weather has often been as warm as in June. Because the month includes the equinox (22 September in 2024), the length of daylight begins to shorten noticeably, and the longer time after sunset allows evening and overnight temperatures to fall. In quiet weather, temperatures soon fall to the saturation point, and dew forms on the ground and vegetation. Temperatures fall even farther overnight, so that

---

Autumn – 10 September to 19 November
This season is marked by an early period when settled weather may occur, as high pressure makes an occasional incursion across the country from the south or south-west. It is, however, usually marked by a period of wet and windy weather, although this tends to die down as the season transitions in the middle of November into that of the early winter.

---

there is often some morning mist and fog. It was in mid-September that John Keats composed his well-loved poem 'To Autumn', with its famous line of 'Season of mists and mellow fruitfulness'.

During the summer, the air arriving from the Atlantic is of the type known to meteorologists as tropical maritime air: cool and fairly humid. The depressions arriving over the country from the Atlantic are fairly shallow, and their winds are subdued. In September, colder air (polar maritime air) begins to come south, and the greater temperature contrast relative to the warmer air from the south tends to deepen depressions and increase their wind speeds.

In Europe, the Full Moon in September is generally known as the 'Harvest Moon'. This is traditionally the Full Moon closest to the equinox. It is also because this is normally the time of the greatest harvest, not only of the various cereal grains, such as wheat and barley, but also apples and similar tree fruits. European harvest festivals were generally held on the Sunday closest to the Full Moon in September.

Stratosphere

The second layer in the atmosphere, lying above the troposphere, in which temperatures either stabilise or begin to increase with height. This increase of temperature is primarily driven by the absorption of solar energy by ozone in the ozone layer. In the lowermost region, between the tropopause and about a height of 20 kilometres, the temperature is stable. Above that there is an overall increase to the top of the stratosphere (the stratopause) at an altitude of about 50 kilometres.

S

Weather Extremes in September

Country	Temp.	Location	Date
Maximum temperature			
England	35.6°C	Bawtry – Hesley Hall (South Yorkshire)	2 Sep 1906
Wales	32.3°C	Hawarden Bridge (Flintshire)	1 Sep 1906
Scotland	32.2°C	Gordon Castle (Moray)	1 Sep 1906
Northern Ireland	27.6°C	Armagh (Co. Armagh)	1 Sep 1906
Minimum temperature			
England	-5.6°C	Stanton Downham (Norfolk) Grendon Underwood (Buckinghamshire)	30 Sep 1969
Wales	-5.5°C	St Harmon (Powys)	19 Sep 1986
Scotland	-6.7°C	Dalwhinnie (Highland)	26 Sep 1942
Northern Ireland	-3.7°C	Katesbridge (Co. Down)	27 Sep 2020

Country	Pressure	Location	Date
Maximum pressure			
Northern Ireland	1042.0 hPa	Ballykelly (Co. Londonderry)	11 Sep 2009
Minimum pressure			
Republic of Ireland	957.1 hPa	Claremorris (Co. Mayo)	21 Sep 1953

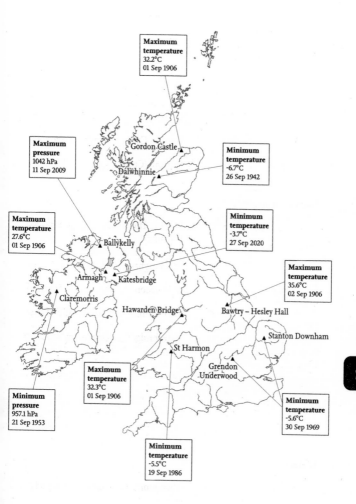

Maximum
temperature
32.2°C
01 Sep 1906

Maximum
pressure
1042 hPa
11 Sep 2009

Gordon Castle

Dalwhinnie

Minimum
temperature
-6.7°C
26 Sep 1942

Maximum
temperature
27.6°C
01 Sep 1906

Ballykelly

Minimum
temperature
-3.7°C
27 Sep 2020

Armagh   Katesbridge

Maximum
temperature
35.6°C
02 Sep 1906

Claremorris

Hawarden Bridge   Bawtry – Hesley Hall

Stanton Downham

St Harmon

Grendon
Underwood

Minimum
pressure
957.1 hPa
21 Sep 1953

Maximum
temperature
32.3°C
01 Sep 1906

Minimum
temperature
-5.6°C
30 Sep 1969

Minimum
temperature
-5.5°C
19 Sep 1986

S

The Weather in September 2022

Observation	Location	Date
Max. temperature 27.7°C	Felsham (Suffolk)	4 September
Min. temperature -1.7°C	Shap (Cumbria)	17 September
Most rainfall 100.4 mm	Seathwaite (Cumbria)	30 September
Highest gust 75 mph (65 Kt)	Needles Old Battery (Isle of Wight)	30 September

Storm surge
A raised level of seawater that is driven ashore and may cause flooding many kilometres inland. The level of the sea is raised primarily by the lower atmospheric pressure at the centre of depressions or tropical cyclones, and may be increased by a high tide (especially a spring tide). The water is often driven ashore by onshore winds. Storm surges have frequently caused many thousands of deaths, especially in vulnerable areas of India and Bangladesh.

September began with a spell of warm weather, but after a few days most regions became fairly unsettled. The morning of the 4th saw localised flooding in Northern Ireland, and over the next few days, rain brought disruption to Taunton, Surrey, Edinburgh and Dundee. The 7th, 8th and 9th saw particularly severe flooding in central and eastern Scotland.

The following part of the month was more settled and, though temperatures fell (Shap in Cumbria recorded an overnight minimum of -1.7°C on the 17th), it was also somewhat drier. Shoeburyness in Essex and Tenby both recorded over 10 hours of bright sunshine on the 17th but temperatures remained cool and mist and fog were common in coastal areas.

From the 22nd onwards, the weather became more broken and autumnal. Isolated showers across England and Scotland contributed to an unsettled picture overall, although sunshine continued to dominate in places. Cloud amounts gradually increased over the rest of the month, with more consistent bands of rain also becoming prevalent.

The final day of the month saw heavy rain and strong winds. The Needles on the Isle of Wight saw gusts of 75 mph, and Fair Isle in Shetland 72 mph. Cumbria was wettest, with 100.4 mm falling at Seathwaite. By the end of the day, however, the winds had broken up the more solid band of rain. In most places this was replaced by showers, occasionally heavy, but parts of Wales and southern England saw fine weather to round out the month.

Overall the month was warmer, wetter and with less sun than average, but there were significant geographical variations within this. North-western Scotland and East Anglia saw less rain than normal, and East Anglia also saw the month's highest temperatures (27.7°C at Felsham in Suffolk on the 4th). Eastern Scotland, Northern Ireland and south-east England were wetter, with some places seeing 150 per cent of their average rainfall. The UK as a whole received 111 per cent, and 92 per cent of its average sunshine. The mean temperature of 13.4°C was 0.5°C above the long-term average.

S

Sunrise and Sunset 2024

Location	Date	Rise	Azimuth °	Set	Azimuth °
Belfast					
	1 Sep (Sun)	05:32	75	19:14	285
	11 Sep (Wed)	05:50	81	18:49	278
	21 Sep (Sat)	06:08	88	18:24	272
	30 Sep (Mon)	06:25	94	18:01	266
Cardiff					
	1 Sep (Sun)	05:26	76	18:58	284
	11 Sep (Wed)	05:42	82	18:35	278
	21 Sep (Sat)	05:58	88	18:12	272
	30 Sep (Mon)	06:12	94	17:52	266
Edinburgh					
	1 Sep (Sun)	05:18	74	19:06	285
	11 Sep (Wed)	05:37	81	18:40	279
	21 Sep (Sat)	05:57	88	18:13	272
	30 Sep (Mon)	06:14	94	17:50	266
London					
	1 Sep (Sun)	05:14	76	18:47	284
	11 Sep (Wed)	05:30	82	18:24	278
	21 Sep (Sat)	05:46	88	18:01	272
	30 Sep (Mon)	06:01	94	17:40	266

Note that all times are in Universal Time (UT), otherwise known as Greenwich Mean Time (GMT). These times do not take Summer Time (BST) into account.

Moonrise and Moonset 2024

Location	Date	Rise	Azimuth °	Set	Azimuth °
Belfast					
	1 Sep (Sun)	03:08	55	19:10	299
	11 Sep (Wed)	15:53	146	21:13	214
	21 Sep (Sat)	19:15	51	11:12	306
	30 Sep (Mon)	03:32	72	17:36	283
Cardiff					
	1 Sep (Sun)	03:11	58	18:48	297
	11 Sep (Wed)	15:13	140	21:41	220
	21 Sep (Sat)	19:20	54	10:45	303
	30 Sep (Mon)	03:27	73	17:20	282
Edinburgh					
	1 Sep (Sun)	02:49	54	19:04	300
	11 Sep (Wed)	15:57	149	20:57	211
	21 Sep (Sat)	18:56	49	11:07	307
	30 Sep (Mon)	03:17	71	17:27	283
London					
	1 Sep (Sun)	02:58	58	18:37	297
	11 Sep (Wed)	15:02	140	21:27	219
	21 Sep (Sat)	19:08	54	10:33	303
	30 Sep (Mon)	03:15	73	17:09	282

S

Note that all times are in Universal Time (UT), otherwise known as Greenwich Mean Time (GMT). These times do not take Summer Time (BST) into account.

Twilight Diagrams 2024

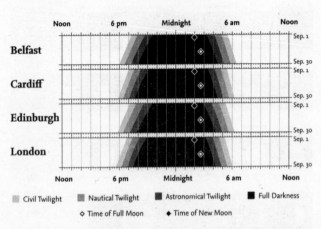

The exact times of the Moon's major phases are shown on the diagrams opposite.

Mesosphere
The third layer of the atmosphere, above the stratosphere and below the thermosphere. It extends from about 50 km (the height of the stratopause) to about 86–100 km (the mesopause). Within it, temperature decreases with increasing altitude, reaching the atmospheric minimum of approximately -123°C at the mesopause. The only clouds occurring within the mesosphere are noctilucent clouds (see pages 112 and 113).

The Moon's Phases and Ages 2024

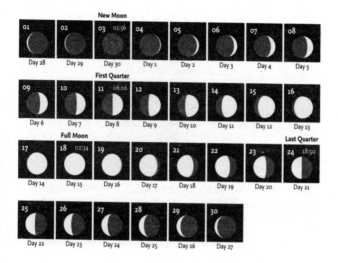

		New Moon					
01	02	03 01:56	04	05	06	07	08
Day 28	Day 29	Day 30	Day 1	Day 2	Day 3	Day 4	Day 5

		First Quarter					
09	10	11 06:06	12	13	14	15	16
Day 6	Day 7	Day 8	Day 9	Day 10	Day 11	Day 12	Day 13

Full Moon							Last Quarter
17	18 02:34	19	20	21	22	23	24 18:50
Day 14	Day 15	Day 16	Day 17	Day 18	Day 19	Day 20	Day 21

25	26	27	28	29	30
Day 22	Day 23	Day 24	Day 25	Day 26	Day 27

Thermosphere
The fourth layer of the atmosphere, counting from the surface.
It is tenuous and lies above the upper limit of the mesosphere,
the mesopause, at approximately 86–100 km, and extends
into interplanetary space. Within it, the temperature increases
continuously with height.

S

September – In this month

1 September 1859 – The Carrington Event, the most significant solar storm on record, caused aurorae to be visible as far south as the Carribean.

2 September 1666 – The Great Fire of London started on Pudding Lane. At the end of a dry summer and fanned by strong north-easterly winds, the fire spread quickly.

5 September 1958 – A severe storm hit Kent, Essex and Sussex. Hail, heavy rain and lightning inflicted flooding and widespread damage to crops. Horsham was particularly badly affected, with hailstones measuring 2 inches in diameter and tornadoes uprooting trees. Electricity was lost for 24 hours and phone lines down for several days in some places.

7 September 1974 – Severe gales led to several deaths, and the stadium roof was blown off during the Derby-Newcastle football match.

7 September 2012 – Construction of the Greater Gabbard wind farm was completed off the Suffolk coast. Including a 2018 extension, the site contains 140 turbines with a combined capacity of 504 MW.

15 September 1968 – Following an already wet start to the month, including a thunderstorm the previous day, more torrential rain caused devastating floods in Surrey. Rain fell continuously for 15 hours, with Tilbury recording as much as 201 mm. Guildford, Lewisham and elsewhere were all affected, but the damage was particularly high in East Molesey. The river Mole burst its banks and several bridges were inundated. The army was called in to assist in evacuation efforts, and residents were instructed to hang a white towel from their window if they required help. Only after several days did the waters begin to subside and people return to their homes to assess the damage.

16 September 1736 – Daniel Fahrenheit died. A physicist, inventor and scientific instrument maker, his legacy included the mercury-in-glass thermometer and the Fahrenheit temperature scale.

18 September 2018 – Storm Ali left homes and businesses without power and affected road, rail and air travel, particularly in Scotland and Northern Ireland. Two people were killed.

27 September 1965 – A landslide in Glen Ogle led to the premature closure of the train line between Dunblane and Crianlarich in Scotland. The line had been scheduled for closure later that year anyway.

S

The Great Fire of London

2–6 September 1666

The Great Fire of London started on Sunday, 2 September 1666. It began with a small fire that started at about two o'clock in the morning, at a baker's shop in Pudding Lane. At the time, London largely consisted of tightly-packed wooden buildings. Many overhung the narrow streets and fires that spread from one building to another were common. There were numerous petitions to various authorities (including to King Charles II) for action, but nothing was done. Basically, the authorities were complacent. A series of wet summers between 1660 and 1664 led to a reduction in the number of fires, which increased this complacency further.

1665, a year in which Britain was hit by bubonic plague, was drier with more fires, but 1666 produced true drought conditions. A succession of anticyclonic weather caused many rivers and streams to dry out completely, and the dry air caused water to evaporate from timber of all sorts, including that in buildings. The timber-framed houses were now bone dry, and thus readily combustible material.

An engraving of London on fire, with the medieval St Paul's cathedral on the horizon. It was later replaced by the current building, designed by Sir Christopher Wren.

Old London Bridge largely consisted of timber buildings, so it was also destroyed by the fire. Many inhabitants took to the River Thames to escape the flames.

During September 1666, London lay underneath the southern edge of a ridge of high pressure that had its centre over Scandinavia and extending across the North Sea. The wind was blowing strongly from the east, possibly increased in strength by a deepening depression to the south, centred over central France. When the fire started, these strong easterly winds fanned the flames and spread the fire westwards. By Monday it became all-too obvious that this was no small fire, but a major conflagration. Although attempts were made to limit the spread by pulling down burning buildings and demolishing others to create firebreaks, the strong winds carried embers across the gaps, which ignited fires on the far side. Despite a slight change in wind direction on Tuesday, it was if anything even stronger. Only when the wind swung round to the south on Wednesday could the blaze be brought under some form of control. The fire was considered 'out' by Thursday, 6 September, although small

S

fires continued to break out from 'hidden' sources (mainly from basements and cellars) for some months afterwards.

Over the five days it burned, the fire destroyed over 435 acres, about a quarter of urban London, including 13,000 wooden houses, 87 churches and the iconic medieval St Paul's Cathedral. Londoners who had fled the fire to the surrounding fields were slow to return, and it took up to 50 years to rebuild the city. The legacy of the fire can be seen today in the plethora of eighteenth-century buildings that were erected in the redesigned streets. The most notable of these are Sir Christopher Wren's classically-inspired churches, including the new St Paul's Cathedral, and The Monument. Also in the aftermath of the fire, insurance schemes began to employ men to put out fires, thus forming the first fire brigades.

October

Introduction

October is definitely an autumnal month. It generally sees an increase in the number of depressions advancing across the country from the Atlantic. These are often vigorous, with strong winds and carrying plenty of rain. The nature of the accompanying weather depends on the type of air within the depressions. Frequently this will include warm, moist maritime tropical air, and give rise to dull days, often accompanied by extensive rain, because the air has passed over a relatively warm sea and thus taken up significant amounts of moisture. The unstable air behind the cold front of depressions often forms showers.

Because the Arctic is now rapidly cooling, any air arriving from the north may be extremely cold. When there is an incursion of frigid maritime Arctic air, the weather may include the first snow, as well as producing significant heavy showers, which may turn thundery and even turn into hailstorms. Scotland tends to experience more thunderstorms at this time, when strong showers arise in the unstable maritime Arctic air behind the cold fronts of depressions, but the air in front of these depressions arises from a warm sea. It may be heavily laden with moisture and give very heavy and prolonged rain ahead of the warm fronts.

Occasionally during the month there may be a quiet anticyclonic period, giving weather that resembles that in September. Generally the month sees the first extensive frosts, except, perhaps, in the very south of England. In Scotland, the trees may lose their leaves in the strong winds without a major display of colour. In England, the winds are often weaker, being farther from the centres of strong depressions, and so the leaves tend to persist on the trees and may provide a brilliant show of colour.

In Europe and Britain, the Full Moon in October was frequently called the 'Hunter's Moon'. This was the time when people prepared for the coming winter, both by hunting game and by slaughtering livestock. Every three years, however, the first Full Moon after the autumnal equinox actually fell in October. It was then customary to call that particular Full Moon the 'Harvest (rather than Hunter's) Moon'. Among the various tribes in North America, there was a tendency to name the October Full Moon to express the idea that it was the time of leaf-fall. Some typical names were 'Leaf-falling Moon', 'Falling Leaves Moon' or 'Fall Moon'. Some names, such as 'White Frost on Grass Moon' expressed the idea that significant frosts had arrived.

Edmond Halley (1656–1742) is primarily known to the general public as an astronomer, and because of the famous comet named after him, the return of which he predicted, but did not live to see. (This was the very first such prediction.) However, he also made contributions to many different fields of science.

As far as meteorology is concerned, Halley made various voyages – he was commissioned into the Royal Navy – to investigate terrestrial magnetism, but these led to fundamental understanding of wind patterns. His voyage to St Helena in the South Atlantic (returning in May 1678) was of particular significance. From data he obtained on that voyage, he eventually published in 1686 a paper and a chart of the wind directions (particularly the trade winds and monsoons) around the world. This was of fundamental importance in establishing the details of the circulation of the atmosphere. He also found the important relationship between barometric pressure and height above sea level.

Edmond Halley (1656–1742), after whom the Antarctic research station is named, in an undated painting by Thomas Murray.

O

Weather Extremes in October

Country	Temp.	Location	Date
Maximum temperature			
England	29.9°C	Gravesend (Kent)	1 Oct 2011
Wales	28.2°C	Hawarden Airport (Flintshire)	1 Oct 2011
Scotland	27.4°C	Tillypronie (Aberdeenshire)	3 Oct 1908
Northern Ireland	24.1°C	Strabane (Co. Tyrone)	10 Oct 1969
Minimum temperature			
England	-10.6°C	Wark (Northumberland)	17 Oct 1993
Wales	-9.4°C	Rhayader, Penvalley (Powys)	26 Oct 1931
Scotland	-11.7°C	Dalwhinnie (Highland)	28 Oct 1948
Northern Ireland	-7.2°C	Lough Navar Forest (Co. Fermanagh)	18 Oct 1993

Country	Pressure	Location	Date
Maximum pressure			
Scotland	1045.6 hPa	Dyce (Aberdeenshire)	31 Oct 1956
Minimum pressure			
Scotland	946.8 hPa	Cawdor Castle (Nairnshire)	14 Oct 1891

Minimum
pressure
946.8 hPa
14 Oct 1891

Maximum
temperature
27.4°C
03 Oct 1908

Minimum
temperature
-11.7°C
28 Oct 1948

Cawdor Castle

Tillypronie ▲ Dyce

Maximum
pressure
1045.6 hPa
31 Oct 1956

Maximum
temperature
24.1°C
10 Oct 1969

Dalwhinnie

Minimum
temperature
-7.2°C
18 Oct 1993

Strabane

Wark

Minimum
temperature
-10.6°C
17 Oct 1993

Lough Navar
Forest

Hawarden Airport

Maximum
temperature
29.9°C
01 Oct 2011

Rhayader, Penvalley

Maximum
temperature
28.2°C
01 Oct 2011

Gravesend

Minimum
temperature
-9.4°C
26 Oct 1931

O

The Weather in October 2022

Observation	Location	Date
Max. temperature 22.9°C	Kew Gardens (Greater London)	29 October
Min. temperature -3.8°C	Aboyne (Aberdeenshire)	15 October
Most rainfall 102.6 mm	Honister Pass (Cumbria)	5 October
Highest gust 110 mph (96 Kt)	Needles Old Battery (Isle of Wight)	31 October

The beginning of October saw numerous weather systems make landfall, prompting a number of warnings. Southern England saw only light showers but further north and west the rain was considerably heavier. Honister Pass in Cumbria saw 102.6 mm of rain on the 5th, while on the same day flooding disrupted railway services between Glasgow and Perth, and roads in mid-Wales were also affected. There was further disruption on the 7th following more heavy rain in Scotland; trains between Perth and Inverness were suspended, while local trains and buses were impacted around Glasgow.

The middle of the month was calmer. Although temperatures dropped, with overnight minima reaching -3.8°C in Aboyne, Aberdeenshire on the 15th and maxima staying 10°C in some places, most of the country was still warmer than average.

The last 11 days of the month saw temperatures rise again and the unsettled weather return, with a series of bands of wet weather moving north across the UK. On the 20th heavy thundery showers caused flooding to roads, railways and property in Kent, Bedford, Essex and Nottinghamshire. Three days later there were reports of localised flooding in the West Midlands, and of one or two small tornadoes in Hampshire. More heavy rain followed on the 28th, with numerous locations in Northern Ireland reporting floods. The month ended with yet more disruption following further heavy rain and high winds (100 mph gusts were recorded at the Needles on the Isle of Wight). Northern Ireland again saw the worst of the rain, with eastern regions less severely affected.

Overall, October was warmer than average and predominantly unsettled. The mean temperature of 11°C was 1.8°C higher than the long-term average, and maximum temperatures in south-east England reached 22.9°C at Kew Gardens on the 29th, nearly 3°C higher than average. The last third of the month was unseasonably warm, with few frosts anywhere. No dry spells lasted longer than a couple of days, but sunshine was above average for most of the country. Rainfall overall was around average, at 115 per cent, but the east coast was drier than normal, and Northern Ireland somewhat wetter. The month concluded the warmest January–October period on record, during which much of East Anglia and south-east England received only half their expected rainfall.

Typhoon

The term used for a tropical cyclone in the western Pacific Ocean. Typhoons are some of the strongest systems encountered anywhere on Earth.

O

Sunrise and Sunset 2024

Location	Date	Rise	Azimuth °	Set	Azimuth °
Belfast					
	1 Oct (Tue)	06:27	95	17:59	265
	11 Oct (Fri)	06:46	101	17:34	258
	21 Oct (Mon)	07:05	108	17:11	252
	31 Oct (Thu)	07:25	114	16:49	246
Cardiff					
	1 Oct (Tue)	06:14	94	17:49	265
	11 Oct (Fri)	06:31	101	17:27	259
	21 Oct (Mon)	06:48	107	17:06	253
	31 Oct (Thu)	07:05	112	16:47	248
Edinburgh					
	1 Oct (Tue)	06:16	95	17:47	265
	11 Oct (Fri)	06:37	102	17:21	258
	21 Oct (Mon)	06:57	108	16:56	251
	31 Oct (Thu)	07:18	115	16:34	245
London					
	1 Oct (Tue)	06:02	94	17:38	265
	11 Oct (Fri)	06:19	101	17:15	259
	21 Oct (Mon)	06:36	107	16:54	253
	31 Oct (Thu)	06:54	112	16:35	248

Note that all times are in Universal Time (UT), otherwise known as Greenwich Mean Time (GMT). These times do not take Summer Time (BST) into account.

Moonrise and Moonset 2024

Location	Date	Rise	Azimuth °	Set	Azimuth °
Belfast					
	1 Oct (Tue)	04:47	82	17:42	273
	11 Oct (Fri)	15:58	157	22:52	225
	21 Oct (Mon)	18:52	35	13:14	325
	31 Oct (Thu)	06:16	109	16:11	247
Cardiff					
	1 Oct (Tue)	04:38	82	17:30	273
	11 Oct (Fri)	15:25	133	23:01	229
	21 Oct (Mon)	19:09	40	12:34	320
	31 Oct (Thu)	05:57	108	16:08	248
Edinburgh					
	1 Oct (Tue)	04:34	82	17:31	273
	11 Oct (Fri)	15:57	139	22:30	223
	21 Oct (Mon)	18:36	32	13:17	328
	31 Oct (Thu)	06:08	110	15:56	246
London					
	1 Oct (Tue)	04:25	82	17:18	273
	11 Oct (Fri)	15:14	133	22:48	229
	21 Oct (Mon)	18:55	40	12:23	320
	31 Oct (Thu)	05:45	108	15:56	248

Note that all times are in Universal Time (UT), otherwise known as Greenwich Mean Time (GMT). These times do not take Summer Time (BST) into account.

O

Twilight Diagrams 2024

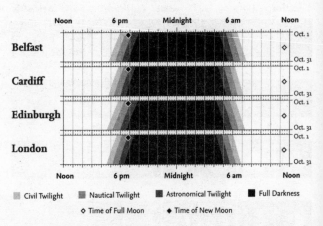

Belfast

Cardiff

Edinburgh

London

| Noon | 6 pm | Midnight | 6 am | Noon |

Civil Twilight   Nautical Twilight   Astronomical Twilight   Full Darkness

◇ Time of Full Moon   ◆ Time of New Moon

The exact times of the Moon's major phases are shown on the diagrams opposite.

Rear-Admiral Sir Francis Beaufort (1774–1857) was a British naval officer who in 1806 devised a means of estimating wind strength at sea. His scheme was not adopted by the Royal Navy until 1838. Beaufort's scale (including an adaptation for use on land) is still in use today (pages 248–251).

In 1829, Beaufort was appointed head of the Admiralty's Hydrographic Office, a post that he held for 25 years. Under his leadership, the Office became the world's leading hydrographic organisation. Beaufort made major contributions to many scientific fields, including geography, geodesy, oceanography and astronomy, as well as meteorology.

The Moon's Phases and Ages 2024

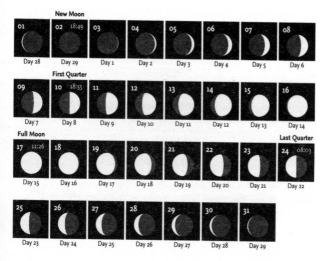

| New Moon | First Quarter | Full Moon | Last Quarter |

| 01 | 02 18:49 | 03 | 04 | 05 | 06 | 07 | 08 |
| Day 28 | Day 29 | Day 1 | Day 2 | Day 3 | Day 4 | Day 5 | Day 6 |

| 09 | 10 18:55 | 11 | 12 | 13 | 14 | 15 | 16 |
| Day 7 | Day 8 | Day 9 | Day 10 | Day 11 | Day 12 | Day 13 | Day 14 |

| 17 11:26 | 18 | 19 | 20 | 21 | 22 | 23 | 24 08:03 |
| Day 15 | Day 16 | Day 17 | Day 18 | Day 19 | Day 20 | Day 21 | Day 22 |

| 25 | 26 | 27 | 28 | 29 | 30 | 31 |
| Day 23 | Day 24 | Day 25 | Day 26 | Day 27 | Day 28 | Day 29 |

Anticyclone
A high-pressure area. Winds circulate around anticyclones in a clockwise direction in the northern hemisphere. (Anticlockwise in the southern hemisphere.) Anticyclones are slow-moving systems (unlike depressions) and tend to extend their influence slowly from an existing centre.

O

October – In this month

1 October 1114 – After a very dry year, rivers were running so low that men could reportedly walk or ride across the Thames east of London Bridge.

1 October 2019 – Torrential rain brought flooding to many parts of Great Britain, with dozens of warnings issued by the Environment Agency. Some areas in the Midlands, Wales and southern England received a week's rain in just one hour.

6 October 2001 – A small tornado swept across the Norfolk Broads, damaging trees, holiday homes and electricity and phone lines. It reportedly carried a column of debris half a mile high.

12 October 1823 – Charles Macintosh, of Scotland, sold the first raincoat. A chemist by training, he pioneered the use of naphtha to make rubber soluble, allowing it to be used to make cotton cloth waterproof. He patented the technique in June of that year, and was soon selling coats made of the new fabric.

15 October 1987 – Michael Fish infamously predicted that there would be 'no hurricane' the following day. While strictly speaking he was correct (the storm system he was referring to did not make landfall), a different weather front brought catastrophic winds, known as the great storm of 1987. Gusts of up to 100 mph brought down 15 million trees and left thousands of homes without power for 24 hours. Eighteen people were killed.

17 October 1091 – A tornado was reported in central London. Based on contemporary accounts, it is thought to have been of strength T8/F4. The resultant river surge washed the old wooden London Bridge away.

22 October 1707 – Four ships sank in the Scilly Naval Disaster. The ships were returning to port as part of a fleet of 15 under the command of Admiral Sir Cloudesley Shovell. The inability to calculate longitude, combined with errors on charts that made latitude readings unreliable, resulted in an error of navigation whereby the fleet was much further north than Shovell assumed. On the evening of the 22nd, several of the ships ran aground on the rocks to the south-west of St Agnes. While some were able to free themselves, four, including the flagship, sank. Admiral Shovell and up to 2,000 other sailors drowned, making it one of the worst disasters in British naval history.

O

The Eruption of Mount Vesuvius

The most recent eruption of Mount Vesuvius was in 1944, when lava destroyed several villages, roads and the bomber aircraft of the US Air Force's 340th Bombardment Group. However, the last major eruption was the famous 2-day eruption in 79 CE, which was described in detail by Roman historians, particularly significant being the two letters written by Pliny the Younger to the historian Tacitus. In these letters he describes, among other things, the final days of his uncle, Pliny the Elder, who died from the effects of the eruption whilst trying to evacuate citizens. The eruption destroyed a number of settlements, the most notable being Pompeii and Herculaneum, but also including the smaller towns of Oplontis, Nuceria and Stabiae.

Perhaps surprisingly, the exact date of the major eruption is not known with certainty, but it was definitely in the autumn, and most probably in late October or early November. A recent study of the two-day eruption has provided dates of 24–25 October, 79 CE.

Although Vesuvius is a stratovolcano – consisting of a series of layers of hardened lava and less consolidated material – most of the destruction came not from lava but from ash fall and pyroclastic flows. Pyroclastic flows are fluidised masses of superheated gas and rock; the rock particles are suspended in the gas, and a flow may travel downhill at speeds of up to 700 kph. Both Pompeii and Herculaneum were buried under many metres of ash, which suffocated and burned the inhabitants with a temperature estimated to have been 250°C. The pyroclastic flows were sufficiently strong to demolish many structures in their path, particularly in Pompeii itself. However, the speed with which the two towns were buried meant that a lot of the buildings and even the outlines of people and animals were preserved extraordinarily well.

Although the ash cloud spread widely, the winds at the time were from the south and south-east so some of the townships, such as Nuceria, were not as deeply buried. Stabiae was also located towards the edge of the ash cloud. Later, smaller eruptions buried the cities of Herculaneum and Pompeii even deeper below volcanic material. According to a traditional tale, Herculaneum was rediscovered in 1709, when a well was being dug. It was definitely known of from 1738, when it was uncovered by workers constructing a summer palace for the King of Naples. For Pompeii, discovery may go as far back as 1592, when an architect was digging an underground aqueduct. However, he kept his discoveries secret. Excavations began in earnest in 1693, when walls were discovered on the site. Pompeii has now been designated a UNESCO World Heritage Site, and the mountain has now become the centre of the Vesuvius National Park.

Vesuvius remains one of the most dangerous volcanoes in the world. Three million people live near enough to be affected by an eruption, and over 600,000 people live in the 'danger zone', where pyroclastic flows pose the highest danger. The Italian Government regularly updates the emergency plan, with the aim of evacuating all inhabitants of the red zone to other parts of the country within 72 hours if there is deemed to be an imminent risk of eruption.

The steep-walled crater of the Vesuvius volcano. The various layers of consolidated lava and of other, finer material are clearly visible.

O

November

Introduction

The weather in the early part of November tends to resemble that in October, so it is definitely an autumnal month. However, morning mists and fogs are much more frequent and also become more persistent, with the decreasing power of the Sun that would otherwise 'burn the fogs off' during the day. Such long-lasting mists and fogs are particularly frequent in the English Midlands and arise either though radiation at night or through the advection of relatively warm, saturated air over cold ground. When such conditions have brought saturated air in an airflow from the west, the descent of the air over the Scottish highlands may result in much warmer temperatures in north-eastern Scotland, particularly in the area around Aberdeen.

Because there is often moderately high pressure over the near Continent at this time of the year, it tends to divert depressions from an easterly track. They then generally veer towards the north, running up the west coast of Britain. The result is heavy rain and strong winds in the west of the country and particularly heavy rainfall (orographic rain) on the mountains of Wales and Scotland. The south and east of England is often much drier

The Full Moon that fell in November, unlike the September 'Harvest Moon' and the October 'Hunter's Moon', did not, in European tradition, have a well-known, universally applied name. It was sometimes known as the 'Frosty Moon' in recognition of the fact that the weather was getting colder, moving towards full winter. On a few occasions we find it called the 'Oak Moon', although this title is more properly applied to the Full Moon in December. Sometimes, if the Full Moon in November was the very last before the winter solstice in December, it was known as the 'Mourning Moon'.

Early winter – 20 November to 19 January

This season typically sees an alternation between long periods of mild westerly weather and drier anticyclonic conditions. Roughly half of the years see this pattern, usually with the westerly episodes being wet and windy, with a succession of depressions arriving from the Atlantic. Short cold periods, generally lasting less than a week, occur in between the westerly episodes. Very cold conditions rarely arrive before the very end of January and generally become established in early February.

N

Weather Extremes in November

Country	Temp.	Location	Date
Maximum temperature			
England	21.1°C	Chelmsford (Essex) Clacton (Essex) Cambridge (Cambridgeshire) Mildenhall (Suffolk)	5 Nov 1938
Wales	22.4°C	Trawsgoed (Ceredigion)	1 Nov 2015
Scotland	20.6°C	Edinburgh Royal Botanic Garden Liberton (Edinburgh)	4 Nov 1946
Northern Ireland	18.5°	Murlough (Co. Down)	3 Nov 1979 1 Nov 2007 10 Nov 2015
Minimum temperature			
England	-16.1°C	Scaleby (Cumbria)	30 Nov 1912
Wales	-18.0°C	Llysdinam (Powys)	28 Nov 2010
Scotland	-23.3°C	Braemar (Aberdeenshire)	14 Nov 1919
Northern Ireland	-12.2°C	Lisburn (Co. Antrim)	15 Nov 1919

Country	Pressure	Location	Date
Maximum pressure			
Scotland	1046.7 hPa	Aviemore (Inverness-shire)	10 Nov 1999
Minimum pressure			
Scotland	939.7 hPa	Monach Lighthouse (Outer Hebrides)	11 Nov 1877

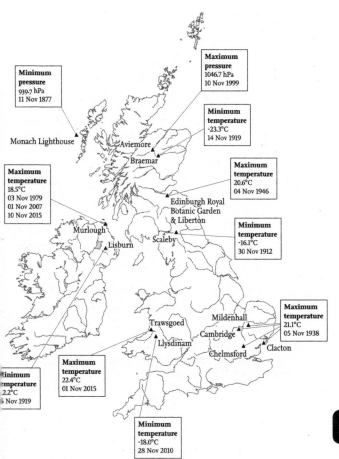

Minimum
pressure
939.7 hPa
11 Nov 1877

Monach Lighthouse

Maximum
pressure
1046.7 hPa
10 Nov 1999

Minimum
temperature
-23.3°C
14 Nov 1919

Aviemore
Braemar

Maximum
temperature
18.5°C
03 Nov 1979
01 Nov 2007
10 Nov 2015

Maximum
temperature
20.6°C
04 Nov 1946

Edinburgh Royal
Botanic Garden
& Liberton

Minimum
temperature
-16.1°C
30 Nov 1912

Murlough
Lisburn
Scaleby

Trawsgoed
Mildenhall
Cambridge
Llysdinam
Chelmsford
Clacton

Maximum
temperature
21.1°C
05 Nov 1938

Minimum
temperature
2.2°C
6 Nov 1919

Maximum
temperature
22.4°C
01 Nov 2015

Minimum
temperature
-18.0°C
28 Nov 2010

N

The Weather in November 2022

Observation	Location	Date
Max. temperature 21.2°C	Porthmadog (Gwynedd)	13 November
Min. temperature -6.0°C	Aviemore (Inverness-shire)	30 November
Most rainfall 115.6 mm	Achnagart (Ross & Cromarty)	11 November
Highest gust 115 mph (100 Kt)	Needles Old Battery (Isle of Wight)	1 November

Picking up from the end of October, the first half of November was very mild and unsettled. The 1st of the month saw heavy rain cause flooding in eastern Scotland, as well as Hampshire and Dorset, while the following day roads were closed in Northern Ireland and rail services were suspended in western Scotland. Strong winds that evening caused a trailer to overturn on the Tay Bridge, and weather disruption continued in London, Essex, Sussex and Kent on the 3rd. There was less rain on the 4th, but more steady rain arrived on the 5th and 6th.

Meanwhile, southerly and south-westerly winds brough warm air from the Azores, culminating in a new record-high minimum for Scotland of 14.6°C at both Kinloss and Prestwick on the night of the 10th. Two days later the highest temperature for the month was set, with 21.2°C recorded at Porthmadog on the 13th. This was accompanied by continued heavy rain, however, particularly in the Highlands and Western Isles.

The second half of the month was cooler, and for a few days it was drier before the remainder of the month was wet again. The 15th to the 18th saw several bands of heavy rain affect south-east England, Sheffield and eastern Scotland. The 19th and 20th saw clearer skies pass from west to east but intense rain on the 21st caused surface water flooding in Pembrokeshire and south-west England. A landslip closed the rail line near Honiton in Devon, and disruption was also reported in Down and Tyrone in Northern Ireland.

The end of the month was slightly drier, and saw the coldest temperatures arrive. The final few days of the month saw mist and fog in eastern regions gradually spread to much of the country. By the final day of the month, fog was lingering all day in some places. Redesdale Camp in Northumberland reached -5.4°C, and Aviemore in Scotland -6°C, on the 30th.

Overall, this was the third-warmest November since 1884, with fewer frosts than average. The UK mean of 8.2°C was 1.8°C above the long-term average, while minimum temperatures were as much as 2.5°C above average in parts of south-east England. It was also a wetter than usual month, with no dry spells lasting longer than three or four days. The UK received 130 per cent of its average rainfall, with the Western Isles, eastern Scotland and southern England particularly wet. Shoreham in West Sussex recorded a total of 224 mm of rain across the month, more than double its average 90 mm, while Polegate in East Sussex saw over 300 mm. Only parts of Scotland and Northern Ireland were noticeably drier than normal.

N

Sunrise and Sunset 2024

Location	Date	Rise	Azimuth °	Set	Azimuth °
Belfast					
	1 Nov (Fri)	07:27	115	16:47	245
	11 Nov (Mon)	07:47	120	16:28	240
	21 Nov (Thu)	08:06	125	16:13	235
	30 Nov (Sat)	08:22	128	16:03	232
Cardiff					
	1 Nov (Fri)	07:07	113	16:45	247
	11 Nov (Mon)	07:25	118	16:28	242
	21 Nov (Thu)	07:41	122	16:16	238
	30 Nov (Sat)	07:55	125	16:08	235
Edinburgh					
	1 Nov (Fri)	07:20	115	16:31	244
	11 Nov (Mon)	07:42	121	16:11	239
	21 Nov (Thu)	08:02	126	15:55	234
	30 Nov (Sat)	08:18	130	15:45	230
London					
	1 Nov (Fri)	06:56	113	16:33	247
	11 Nov (Mon)	07:13	118	16:16	242
	21 Nov (Thu)	07:30	122	16:03	238
	30 Nov (Sat)	07:44	125	15:55	235

Note that all times are in Universal Time (UT), otherwise known as Greenwich Mean Time (GMT). These times do not take Summer Time (BST) into account.

Moonrise and Moonset 2024

Location	Date	Rise	Azimuth °	Set	Azimuth °
Belfast					
	1 Nov (Fri)	07:33	119	16:20	237
	11 Nov (Mon)	14:56	99	01:08	254
	21 Nov (Thu)	21:43	55	13:34	308
	30 Nov (Sat)	07:57	134	14:56	224
Cardiff					
	1 Nov (Fri)	07:09	117	16:21	240
	11 Nov (Mon)	14:41	98	01:02	256
	21 Nov (Thu)	21:46	58	13:07	305
	30 Nov (Sat)	07:26	130	15:06	228
Edinburgh					
	1 Nov (Fri)	07:27	120	16:03	236
	11 Nov (Mon)	14:56	99	00:53	254
	21 Nov (Thu)	21:24	54	13:30	309
	30 Nov (Sat)	07:55	136	14:35	222
London					
	1 Nov (Fri)	06:57	117	16:09	240
	11 Nov (Mon)	14:30	98	00:49	255
	21 Nov (Thu)	21:23	58	12:56	305
	30 Nov (Sat)	07:14	130	14:53	228

N

Note that all times are in Universal Time (UT), otherwise known as Greenwich Mean Time (GMT). These times do not take Summer Time (BST) into account.

Twilight Diagrams 2024

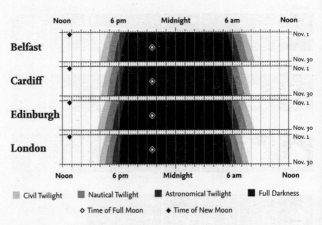

The exact times of the Moon's major phases are shown on the diagrams opposite.

Sea breeze

A flow of air from the sea onto the land. The land heats more rapidly than the sea, so the air above it rises, drawing cooler air off the sea. There is a corresponding flow towards the sea at altitude. The air rises along a 'sea-breeze front', which may lie many kilometres inland, depending on the local geography.

The Moon's Phases and Ages 2024

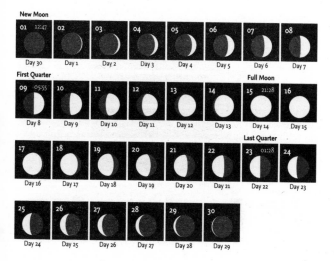

New Moon							
01 12:47	02	03	04	05	06	07	08
Day 30	Day 1	Day 2	Day 3	Day 4	Day 5	Day 6	Day 7
First Quarter						**Full Moon**	
09 05:55	10	11	12	13	14	15 21:28	16
Day 8	Day 9	Day 10	Day 11	Day 12	Day 13	Day 14	Day 15
						Last Quarter	
17	18	19	20	21	22	23 01:28	24
Day 16	Day 17	Day 18	Day 19	Day 20	Day 21	Day 22	Day 23
25	26	27	28	29	30		
Day 24	Day 25	Day 26	Day 27	Day 28	Day 29		

Land breeze
A flow of air from the land towards the sea. At night, the land cools more quickly than the sea. The denser air flows out towards the sea. A 'land-breeze front' is sometimes marked (especially on satellite images) by a line of cumulus cloud, where the air rises to flow back towards the land at altitude.

N

November – In this month

1 November 1965 – Five of the eight cooling towers at the uncompleted Ferrybridge C power station in West Yorkshire collapsed in high winds.

2 November 1925 – The Eigiau Dam in Conwy Valley failed following two weeks of heavy rain, and the resultant floodwaters caused another dam downstream to collapse as well. A total of 70 billion gallons of water and debris were sent cascading into the village of Dolgarrog, killing ten adults and six children. The dam, owned by the Aluminium Corporation, was found at the inquest to have had inadequate foundations.

6 November 2018 – Renewable energy capacity overtook that of fossil fuels in the UK for the first time, at 41.9 GW.

10 November 2015 – Storm Abigail was the first storm to be named by the Met Office. The new naming system is intended to improve the communication of storm preparedness via the media, better equipping the public and business to keep themselves safe.

11 November 1570 – A series of storms began that would last until the 22nd, known as the All Saints Flood. In the Netherlands, dykes gave way on the 21st, drowning an estimated 100,000 people.

15 November 1928 – The Mary Stanford lifeboat capsized while on service in Rye Harbour. The boat had been launched in a south-westerly gale and heavy rain to assist another vessel, which was in the meantime rescued by another boat. The crew failed to see the recall signal and the boat capsized as they eventually returned to the harbour. All 17 crew were lost.

17 November 1852 – The Duke of Wellington's hearse was upset at Bath Road in Maidenhead during one of the highest Thames floods on record, which became known as the Duke of Wellington's Flood. The three weeks that followed brought more rain and flooding, with the Fens and Somerset Levels both inundated.

23 November 1981 – A record 104 tornadoes touched down in Britain in one day. Damage included roofs blown off, trees uprooted and a caravan blown into a lake in Northamptonshire.

27 November 1703 – The first Eddystone Lighthouse was destroyed in the Great Storm of 1703, just five years after completion. Several successive lighthouses would be built at the same location. The current one was built in 1882, and was the first Trinity House rock lighthouse to be converted to fully automatic operation.

N

The Eddystone Lighthouse

One of the most famous lighthouses in the world is that on the Eddystone rock, south-west of Plymouth – and thus a serious obstacle to all ships approaching the harbour from the English Channel.

The original lighthouse on the Eddystone reef was constructed in 1698 by Henry Winstanley, prompted by the loss of his second ship, the Constant, on the rock. Winstanley went on to construct a higher, far more ornate, structure, in 1699. The light was an immediate success and no vessel was lost on the reef whilst it was in operation.

Winstanley was understandably proud of his achievement, and expressed the wish to be present 'during the greatest storm'. He was actually present at the lighthouse during the highly destructive Great Storm of 26 November 1703 OS (Old Style, or 7 December NS – New Style), when the lighthouse

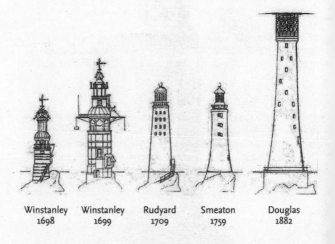

| Winstanley | Winstanley | Rudyard | Smeaton | Douglas |
| 1698 | 1699 | 1709 | 1759 | 1882 |

Artist's impression of the five lighthouses built on the Eddystone Reef since 1698, including (far right) the present-day structure incorporating the helicopter landing pad constructed in 1980.

was completely destroyed and Winstanley killed. (The Great Storm of 26 November 1703 was probably the strongest ever experienced in Great Britain.) Almost immediately, ships were lost on the reef.

Another lighthouse was erected by Rudyard in 1709, but was of wooden construction, and burnt down in 1756. The civil engineer, John Smeaton, then erected a stone tower, of interlocking (dovetailed) granite blocks, in 1759, This was so

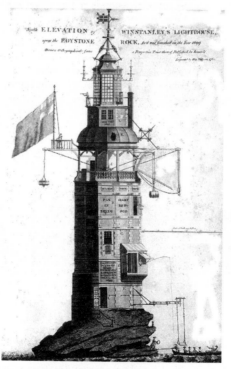

A design sketch of Winstanley's second lighthouse, recently used as the basis for a fanciful modern reconstruction

N

successful that it was finally dismantled in 1887, when the rock on which it was constructed began to be eroded, it was re-erected by public subscription on Plymouth Hoe, where it may still be seen and is known as 'Smeaton's Tower'. Smeaton's lighthouse was not high enough to avoid being overtopped by extreme waves, so the current tower, by James Douglas, erected in 1882, is nearly twice the height (see below).

The modern lighthouse, built by James Douglas, and topped by a helicopter landing pad, alongside the stump of Smeaton's lighthouse.

December

Introduction

For meteorologists, December is the beginning of winter. However, based on the actual weather that occurs during the month, it must be linked with late November and January, because, certainly nowadays, it sees very little severe weather. It may be – and often is – very windy with some notable storms during the month, but it is rarely very cold. The persistent idea of a white, snowy Christmas is actually a hangover from the writings of Charles Dickens in particular. In his young days, early in the nineteenth century, the weather around the time of Christmas was indeed more severe. Today, December is rarely as cold as January or February and on average, the month sees just two days when snow is lying.

December may see the winter solstice (21 December in 2024), and the shortest day, but the temperature remains mild, mitigated by the sea, which is still relatively warm. As a whole then, December is marked by short days, accompanied by wet and windy weather. There have been some notably violent storms that have arrived in December, such as Storm

---

Hadley cell
One of the two atmospheric circulation cells that are driven by the hot air that rises in the tropics (i.e., along the equator or the heat equator that moves north or south with the seasons). The descending limbs are located over the sub-tropical highs at about latitudes 30° north and south.

Desmond and Storm Eva in December 2015. The rainfall from Storm Desmond broke the United Kingdom's 24-hour rainfall record, with 341.4 mm falling at Honister Pass, Cumbria, on December 5. Rain from Storm Eva added to the problems posed by Storm Desmond, and the precipitation caused severe flooding in Cumbria, where the towns of Appleby, Keswick and Kendal were all flooded on 22 December. Some smaller locations were flooded three times during the month.

December is associated with Yule and Yuletide, originally a Germanic festival around the time of the winter solstice. The original pagan festival was taken over by the Christian church, and became Christmas. The Full Moon in December does not have a widely recognised name, unlike the Harvest and Hunter's Moons in September and October, but in the European tradition was sometimes known as 'the Moon before Yule' or even, occasionally, as the 'Wolf Moon', although that name is normally associated with the Full Moon of January.

Ferrel cell
One of the two atmospheric circulation cells that are the intermediate cells, between the Hadley cells, closest to the equator, and the polar cells in each hemisphere. The air, spreading out towards the poles from the sub-tropical highs is diverted towards the east by the rotation of the Earth, and forms the dominant westerlies that govern the weather in the middle latitudes.

D

Weather Extremes in December

Country	Temp.	Location	Date
Maximum temperature			
England	17.7°C	Chivenor (Devon)	2 Dec 1985
		Penkridge (Staffordshire)	11 Dec 1994
Wales	18.0°C	Aber (Gwynedd)	18 Dec 1972
Scotland	18.7°C	Achafry (Sutherland)	28 Dec 2019
Northern Ireland	16.7°C	Ballykelly (Co. Down) (Co. Londonderry)	2 Dec 1948
Minimum temperature			
England	-25.2°C	Shawbury (Shropshire)	13 Dec 1981
Wales	-22.7°C	Corwen (Denbighshire)	13 Dec 1981
Scotland	-27.2°C	Altnaharra (Highland)	30 Dec 1995
Northern Ireland	-18.7°C	Castlederg (Co. Tyrone)	24 Dec 2010

Country	Pressure	Location	Date
Maximum pressure			
Scotland	1051.9 hPa	Wick (Caithness)	24 Dec 1926
Minimum pressure			
Northern Ireland	927.2 hPa	Belfast (Co. Antrim)	8 Dec 1886

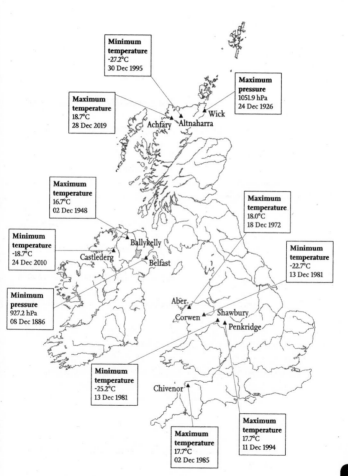

Minimum
temperature
-27.2°C
30 Dec 1995

Maximum
pressure
1051.9 hPa
24 Dec 1926

Maximum
temperature
18.7°C
28 Dec 2019

Wick

Achfary   Altnaharra

Maximum
temperature
16.7°C
02 Dec 1948

Maximum
temperature
18.0°C
18 Dec 1972

Minimum
temperature
-18.7°C
24 Dec 2010

Ballykelly

Castlederg

Belfast

Minimum
temperature
-22.7°C
13 Dec 1981

Minimum
pressure
927.2 hPa
08 Dec 1886

Aber

Corwen   Shawbury

Penkridge

Minimum
temperature
-25.2°C
13 Dec 1981

Chivenor

Maximum
temperature
17.7°C
02 Dec 1985

Maximum
temperature
17.7°C
11 Dec 1994

D

199

The Weather in December 2022

Observation	Location	Date
Max. temperature 15.9°C	Prestatyn (Glwyd) and Hawarden (Glwyd)	19 December
Min. temperature -17.3°C	Braemar (Aberdeenshire)	13 December
Most rainfall 150.6 mm	White Barry (Devon)	19 December
Highest gust 82 mph (71 Kt)	Capel Curig (Gwynedd)	19 December
Greatest snow depth 15 cm	Loch Glascarnoch (Ross & Cromarty)	15 December

The first half of December was cold, with a pronounced cold spell between the 6th and 17th seeing snow and ice. After some wintry showers across Devon and Cornwall on the 10th, the 11th saw heavy snowfall over London and south-east England. Parts of Kent saw more than 10 cm of fresh snow, and the M25 was closed for several hours in Essex. Runways at Gatwick and Stansted were temporarily closed to allow snow to be cleared.

Further north, Scotland saw heavy snow on the 12th, causing damage to electricity cables on Shetland. A minimum temperature of -17.3°C was recorded at Braemar, Aberdeenshire on the 13th. The previous day it had seen the lowest daily maximum of the month, at only -9.3°C. By this point in the month, the freezing temperatures were well established across the country. The Cambridgeshire Fens, with their shallow flooded fields, froze solid, and a revival of Fen skating was seen for the first time in several years. More heavy snow fell across central Scotland on the 15th and 16th, with a snow depth of 15 cm recorded at Loch Glascarnoch. Numerous roads were closed and Glasgow airport suffered delays and cancellations.

The cold spell finally relented over the weekend of the 17th and 18th. The temperature rise was rapid enough to cause several burst water pipes and mains, particularly in Wales and Gloucestershire. The milder weather also brought more precipitation, with showers frequent for the rest of the month. White Barrow in Devon received 150.6 mm of rain on the 19th.

Christmas Day was wet in most places and the final few days of the month saw prolonged periods of rain across all four nations. This caused disruption in Scotland and Northern Ireland, with landslips closing the West Coast Main Line between Lockerbie and Carstairs, properties flooded in Dumfries after the River Nith overtopped its banks, and roads closed in County Down.

In contrast to the preceding months, December 2022 was colder than average overall, with a mean temperature of 2.9°C. The very cold and frosty spell spanning the first half of the month, during which minimum temperatures fell around 2°C below average in some parts of England and Wales, was only partially offset by milder weather afterwards. Although there was limited rain during the cold snap, heavy rain towards the end of the month brought the total amount close to the long-term average at 87 per cent. Sunshine was slightly above average overall, and particularly so in the far north and east where some places received 116 per cent of average sunshine.

Exosphere
The name sometimes applied to the upper region of the thermosphere above an altitude of 200–700 km.

D

Sunrise and Sunset 2024

Location	Date	Rise	Azimuth °	Set	Azimuth °
Belfast					
	1 Dec (Sun)	08:23	129	16:02	231
	11 Dec (Wed)	08:37	131	15:58	229
	21 Dec (Sat)	08:45	132	16:00	228
	31 Dec (Tue)	08:46	131	16:08	229
Cardiff					
	1 Dec (Sun)	07:57	125	16:07	234
	11 Dec (Wed)	08:08	128	16:04	232
	21 Dec (Sat)	08:16	128	16:06	232
	31 Dec (Tue)	08:18	128	16:14	232
Edinburgh					
	1 Dec (Sun)	08:20	130	15:44	230
	11 Dec (Wed)	08:34	133	15:38	227
	21 Dec (Sat)	08:42	134	15:40	226
	31 Dec (Tue)	08:44	133	15:49	227
London					
	1 Dec (Sun)	07:45	126	15:55	234
	11 Dec (Wed)	07:57	128	15:51	232
	21 Dec (Sat)	08:05	128	15:54	232
	31 Dec (Tue)	08:07	128	16:01	232

Note that all times are in Universal Time (UT), otherwise known as Greenwich Mean Time (GMT). These times do not take Summer Time (BST) into account.

Moonrise and Moonset 2024

Location	Date	Rise	Azimuth °	Set	Azimuth °
Belfast					
	1 Dec (Sun)	09:17	141	15:22	218
	11 Dec (Wed)	13:29	67	03:21	288
	21 Dec (Sat)	23:18	82	12:09	283
	31 Dec (Tue)	10:02	143	16:07	219
Cardiff					
	1 Dec (Sun)	08:41	136	15:36	223
	11 Dec (Wed)	13:27	69	03:02	286
	21 Dec (Sat)	23:10	82	11:53	282
	31 Dec (Tue)	09:26	138	16:21	224
Edinburgh					
	1 Dec (Sun)	09:18	144	14:58	215
	11 Dec (Wed)	13:14	67	03:12	288
	21 Dec (Sat)	23:05	81	12:00	283
	31 Dec (Tue)	10:04	145	15:43	216
London					
	1 Dec (Sun)	08:29	136	15:23	222
	11 Dec (Wed)	13:14	138	02:50	286
	21 Dec (Sat)	22:57	82	11:42	282
	31 Dec (Tue)	09:14	138	16:07	223

Note that all times are in Universal Time (UT), otherwise known as Greenwich Mean Time (GMT). These times do not take Summer Time (BST) into account.

D

Twilight Diagrams 2024

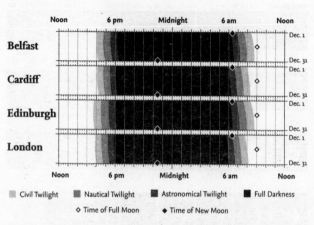

The exact times of the Moon's major phases are shown on the diagrams opposite.

Polar cell

One of the two cells, farthest from the equator, where cold air spreads out from the poles, heading in the general direction of the equator. The winds are easterly. This air meets the air in the Ferrell cells at the polar fronts, which tend to vary in latitude, and which are where the all-important depressions form.

The Moon's Phases and Ages 2024

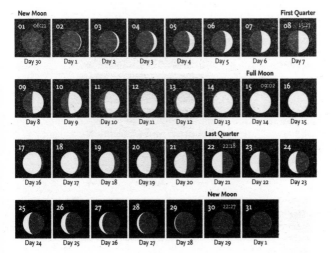

New Moon

							First Quarter
01 06:21	02	03	04	05	06	07	08 15:27
Day 30	Day 1	Day 2	Day 3	Day 4	Day 5	Day 6	Day 7

Full Moon

09	10	11	12	13	14	15 09:02	16
Day 8	Day 9	Day 10	Day 11	Day 12	Day 13	Day 14	Day 15

Last Quarter

17	18	19	20	21	22 22:18	23	24
Day 16	Day 17	Day 18	Day 19	Day 20	Day 21	Day 22	Day 23

New Moon

25	26	27	28	29	30 22:27	31	
Day 24	Day 25	Day 26	Day 27	Day 28	Day 29	Day 1	

Polar front
The area where cold air from the poles meets warmer air that has spread out from the sub-tropical high-pressure regions. The conflict between the two types of air creates depressions which bring most of the changeable weather to countries in the middle latitudes.

D

December – In this month

1 December 2010 – Heavy snow and freezing temperatures caused widespread transport closure, including Edinburgh and Gatwick airports. More airports shut over the following month, with Heathrow shut for four days and Christmas travel plans severely disrupted.

5 December 1952 – A cold fog descended on London, combining with air pollution to cause the Great Smog of 1952. At least 12,000 people died from the smog's effects over the following weeks and months.

9 December 2021 – Longannet power station, the last remaining coal-fired power station in Scotland, was demolished.

14 December 1287 – A North Sea surge known as St Lucia's Flood caused widespread coastal damage. Much of the port of Dunwich was destroyed, and the city never recovered. What had been one of the largest and most important cities in the country, rivalling London and Norwich for economic significance, was gradually abandoned to erosion over the next few hundred years.

24 December 1839 – The biggest recorded landslip in British history occurred between Lyme Regis and Axmouth, following several months of rain that saturated the cliffs. In total, 50 acres of land slipped to the beach, and this portion of the Jurassic coast is now known as the Undercliff.

26 December 1998 – A severe storm in southern Scotland led to the complete closure of the Forth Road Bridge for the first time ever.

27 December 1813 – A thick fog blanketed London, making travel all but impossible. The Prince Regent was forced to turn back from a trip to Hatfield House, while the Birmingham mail coach took seven hours to reach Uxbridge.

27 December 1836 – The deadliest avalanche on record in the UK occurred at Lewes, Sussex, when a build-up of snow fell 100 m from a cliff onto the town. A row of houses was destroyed and, while a rescue operation succeeded in saving seven people, eight were killed.

28 December 1879 – The Tay rail bridge collapsed in a violent storm while a train was crossing, resulting in the deaths of all on board. It was found that no allowance had been made for wind loading in the design of the bridge, and its designer, Thomas Bouch, saw his reputation ruined.

D

I'm Dreaming of a White Christmas...

So runs the song, and a snow-bound scene is certainly a central aspect of today's commercial image of Christmas, featuring in every other card, shop window display and chocolate tin. But how common is such an occurrence in the UK – and was it ever the norm?

For official betting purposes, the Met Office defines a white Christmas as 'one snowflake to be observed falling in the 24 hours of 25 December somewhere in the UK, by which measure over half of all years on record have been white Christmases. However, a more traditional metric of snow on the ground is far less common. For instance, 2021 was technically a white Christmas, with 6 per cent of weather stations recording snow falling, but fewer than 1 per cent reported any snow lying on the ground. There has only been widespread snow on the ground (i.e. at more than 40 per cent of weather stations) four times since 1960. For places in the south of England, the frequency of seeing snow lying on Christmas day is around 1 in

Mr Pickwick sliding on the ice. From The Pickwick Papers, *by Charles Dickens. Illustration by Hablot Knight Browne.*

25 years, while for the north of Scotland it might be as common as 1 in 10 years.

White Christmases were indeed more common during the so-called Little Ice Age, which spanned from the 1550s to the 1850s. This was the period when 'frost fairs' were held on the frozen Thames roughly once in a generation, and the consistently colder temperatures meant that any snow in the fortnight or so before the 25th was likely to mean snow on the ground on Christmas morning. Thus the scenes in Charles Dickens' Pickwick Papers, A Christmas Carol and other Victorian novels were describing an at least semi-familiar phenomenon, but by the end of the nineteenth century it was already becoming less common. Exacerbating the unlikelihood further was the shift from the Julian to the Gregorian calendar in 1752. This effectively moved Christmas 12 days earlier in the winter season, whereas most snow in the UK tends to fall later, in what is now January and February.

The last widespread white Christmas in the UK was in 2010, when 83 per cent of weather stations reported snow on the ground. The month of December was the UK's coldest for 100 years, with snow falling throughout the month and depths reaching over 50 cm in places. Christmas travel plans were severely disrupted across Europe. London Heathrow was all but closed for three days from the 19th, and roads and railways across the country continued to be affected through Boxing Day. Although the worst of the snow abated in time for people in the UK to reach their families for Christmas, elsewhere in Europe this was not the case; passengers were still stranded at Charles de Gaulle airport in Paris on Christmas morning.

With human-driven climate change causing temperatures to rise year on year, the chance of snow at Christmas continues to decrease. The number of snowy days in London in December dropped by three quarters between 1970 and 2020, to an average of just one snowy day every three years. Northern Scotland is the snowiest part of the UK, but there too the number of snowy days has dropped, from 6.1 per year in 1970 to 4.2 per year in 2020. If you want to take that postcard-perfect picture, you might be waiting a while.

D

Additional
Information

The Regional Climates of Britain

1   South-West England and the Channel Islands

The south-western region may be taken to include Cornwall, Devon, Somerset, Gloucestershire, Dorset and the western portion of Wiltshire. This area is largely dominated by its proximity to the sea, although the northern and eastern portion of the region, being farther from the sea, often experiences rather different weather. In many respects the closeness of the Atlantic means that the weather resembles that encountered in the west of Ireland or the Hebrides. Generally, the climate is extremely mild, although that in the Scilly Isles is drier, sunnier and much milder than the closest part of the Cornish peninsula, just 40 km farther north. In the prevailing moist, south-westerly airstreams, the Scilly Isles are not only surrounded by the sea, but they are fairly flat with no hills to cause the air to rise and produce rain. The Channel Islands, by contrast, well to the east, are affected by their proximity to France and sometimes come under the influence of anticyclonic high-pressure conditions on the near Continent, so their overall climate tends to be more extreme.

Despite its generally mild weather, the region has experienced extremes, such as the exceptional snowfall in March 1891 that paralysed southern counties and introduced the word 'blizzard' to descriptions of British weather. The region also experienced the British rainstorm record of 279 mm in one single observational day (09.00 am one day to 08.59 am the next) occurring on 18 July 1955 at Martinstown in Dorset.

Most of the peninsula of Devon and Cornwall sees very few days of frost and some areas are almost completely frost-free. Temperatures are lower, of course, over the high ground of Bodmin Moor and Dartmoor. Indeed, those areas and the Mendip Hills and Blackdown Hills do all have slightly different climatic regimes. The influence of the Severn Estuary extends well inland, and actually has an effect on the weather in the Midlands (see page 218). In winter, it allows mild air to penetrate far inland. In Cornwall and Devon, particularly in summer, sea breezes from opposite sides of the peninsula converge over the high ground that runs along the centre of the peninsula, leading to the formation of major cumulonimbus

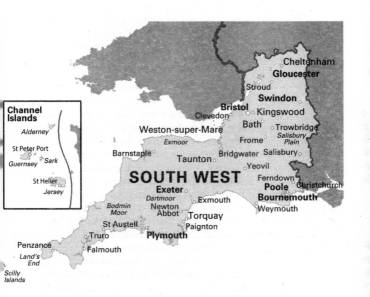

clouds and frequent showers, which may give extreme rainfall. It was this that led to the Lynmouth disaster in August 1952, when waters from a flash flood devastated the town and caused the deaths of 34 people. A somewhat similar situation arose in August 2004 in nearby Boscastle and Crackington Haven, although in that instance, no lives were lost.

The escarpment of the Cotswolds, overlooking the Severn valley, often proves to be a boundary between different types of weather. This is particularly the case when there is a north-westerly wind. Then heavy showers may affect areas on the high ground, while it is warmer and with less wind over the flatter land in the Severn valley and around Gloucester. It is often bitterly cold on the high ground above the escarpment.

2 South-East England and East Anglia

The weather in the south-eastern corner of the country may be divided into two main areas: the counties along the south coast (Hampshire, West and East Sussex and Kent); and the Home Counties around London, and East Anglia, although East Anglia (Norfolk and Suffolk in particular) often experiences rather different conditions to the Home Counties.

The coastal strip from Hampshire (including the Isle of Wight) eastwards to southern Kent has long been recognised as the warmest and sunniest part of the British Isles. This largely arises from the longer duration of warm tropical air from the Continent when compared with the length of time that such air penetrates to more northern areas. The coastal strip from Norfolk to northern Kent does experience some warming effect when the winds are in the prevailing south-westerly direction, offsetting the cold effects produced by the North Sea. This coast may experience severe weather when there is an easterly or north-easterly airflow over the North Sea. This is particularly the case in winter: cold easterly winds bring significant snowfall to the region. It is also a feature of spring and early summer when temperatures are reduced when there is an onshore wind off the North Sea.

Frosts and frost hollows are a feature of the South and North Downs, the Chiltern Hills (in Berkshire, Bedfordshire and

Hertfordshire), and in the high ground in East Anglia. Here, the chalk subsoil loses heat by night as do the sandy soils of Surrey and Breckland in Norfolk, leading to ground frost in places in any month of the year.

Along the south coast there is a tendency for most rain to fall in the autumn and winter, whereas for the rest of the region (the Home Counties and East Anglia) precipitation tends to occur more-or-less equally throughout the year. Because of the reliance upon groundwater, the whole region sometimes suffers from drought, when winter rains (in particular) have been insufficient to recharge the underground reserves.

3 The Midlands

The Midlands region consists of a very large number of counties: Shropshire, Herefordshire, Worcestershire, Warwickshire, West Midlands, Staffordshire, Nottinghamshire, Lincolnshire, Leicestershire, Rutland, Northamptonshire and the southern part of Derbyshire (excluding the high ground of the High Peak in the north). Of all the regions of the British Isles, this is naturally the area which has the least maritime influence. The region has been likened to a shallow bowl surrounded by hills (the Welsh Marches, the Cotswolds, the Northamptonshire Escarpment, the Derbyshire Peak and the Staffordshire Moorlands) and with a slight dome in the centre (the Birmingham Plateau). In winter the warmest area is that closest to the Severn Valley, where warm south-westerly winds may penetrate inland, whereas in summer the warmest region lies to the north-east, farthest from those moderating winds. Yet the western area is also prone to very cold nights in autumn, winter and early spring. (The lowest temperature ever recorded in England was -26.1°C at Newport in Shropshire on 10 January 1982.) Frosts are a feature of the whole region, partly because of the sandy nature of most of the soils and also because of the lack of maritime influence.

Precipitation is fairly evenly spread across the region, although the west, along the Welsh border and the high northern area of the High Peak, experiences the highest rainfall. The east (Lincolnshire and the low ground in the

east of Nottinghamshire and Northamptonshire along the valleys of the Trent and Nene), tends to be drier. Because of the rain-shadow created by the Welsh mountains, over which considerable rainfall occurs, some areas of the west of this Midlands region are drier than might otherwise be expected. There is some increase in rainfall over the slightly higher ground of the Birmingham Plateau and also towards the south and the Cotswold hills. Towards East Anglia there is a strong tendency for most rain to occur in summer, when showers are most numerous. In the very hilly areas on the Welsh border and in the Peak District, the wettest months are December and January. The Derbyshire Peak and the Staffordshire Moorlands tend to experience considerable snowfall in winter, as do high areas of the Welsh Marches. On the lower ground to the east, snowfall is greater in the year than in the west. This is particularly the case when there are easterly or north-easterly winds that penetrate inland and bring snow from the North Sea.

4 North-West England and the Isle of Man

The North West region consists of the land west of the Pennine chain, that is Cumbria and Lancashire in particular, especially including the mountainous Lake District in Cumbria, but also extends south to include Merseyside, Greater Manchester, Cheshire and the western side of Derbyshire. The region's weather tends to be mild and wetter in winter than regions to the east of the Pennines, and cooler in summer than regions to the south. It was, of course, the mild, relatively damp climate that was responsible for the region being the centre for the spinning of cotton, in contrast to the wool handled in the drier east. The maritime influence is seen in the fact that coastal areas are often warmer in winter and cooler in summer than areas farther inland.

There is a great difference in the amount of precipitation between the north and south of this region. The north, in Cumbria and the Lake District is notorious for high rainfall. Seathwaite, in the Lake District, currently holds the record for rainfall in 24 hours and is the wettest inhabited location

in Britain. The extreme rainfall has often contributed to severe flooding, such as that in 2005 and 2009 in Carlisle, Cockermouth, Workington, Appleby and Keswick. By contrast, rainfall is much less over the Cheshire plain, which like much of Merseyside, actually lies in the rain shadow of the Welsh mountains and is thus much drier.

The prevailing wind from the south-west may give very high wind speeds over the high ground of the Pennines, while an easterly wind may produce the only named British wind, the viciously strong, and noisy, Helm wind, as air cascades over the escarpment west of Cross Fell and over the Eden valley in the north of Cumbria.

Most years see some early snow in autumn on the high fells, and in the north on the high ground it may be persistent, although it rarely lasts throughout the winter. The low ground along the coast and in the south sees relatively little snow and what does fall remains lying for just a few days.

5 North-East England and Yorkshire

The region of North East England is well-defined by two geographical boundaries. On the west is the Pennine range, on the east, the North Sea. The northern boundary may be taken as the river Tweed and the southern as the estuary of the Humber. The region includes very high moorland in the north and west. In general the ground slopes down from the Pennine chain towards the east coast. There are, however, considerable areas of lower land, such as that in parts of Northumberland and, in particular, in the south of Yorkshire. Because the prevailing winds in Britain are from the west, they tend to deposit most of their rainfall over the Pennines, so that there is a rain shadow effect that reaches right across this region to the coast, and the whole region is drier than might otherwise be expected. With westerly winds the high fells also tend to break up the cloud cover, so that the whole area to the east is surprisingly sunny. On the other hand, the North Sea exerts a strong cooling effect, keeping general temperatures fairly low. The region is, however, open to easterly and northerly winds and tends to suffer from gales off

the sea, accompanied by heavy rain or, in winter, by snow. The North Sea never becomes particularly warm and exerts a chilling effect over the whole length of the region. Sea breezes, which occur because of the difference between the warm land and the cold sea, commonly set in, especially in late spring and often bring damp sea fog, or 'haar' to the coastal strip. Such conditions frequently arise when other parts of the country, especially the south-east, are enjoying settled, warm, anticyclonic weather. The cooling effect of the North Sea may be so great that there is a difference of as much as 10 degrees Celsius between the coastal strip and locations just a few kilometres inland. This difference tends to be greatest during the summer months.

Although the region has a low overall rainfall because of the effect of the Pennines to the west, when the wind is easterly it may produce prolonged heavy rainfall, as the air is forced to rise over the high ground, causing it to deposit its moisture as rain or, in winter, as snow. Long periods of heavy rain may occur from the easterly airflow on the northern side of depressions, the centres of which are tracking farther south across the country. Most of the rain is of this, frontal, type as there are few of the convective showers that produce heavy rainfall in regions farther south. Most of the rivers in the region have large catchment areas, extending well into the Pennines, and are therefore subject to episodes of flooding when there is prolonged rain. This is particularly the case in the south of the region, where the Ouse frequently floods around York and where the city of Hull often suffers.

6 Wales

Although Wales is being treated here as a single climatic region, in reality there are considerable differences in various areas. The whole country, is, of course, dominated by the long mountainous spine running from north to south. This area is not only high, but it is also exposed, windy and wet. To the west, the area around the whole of Cardigan Bay has a very

maritime climate. It is windy, but may be particularly fine when the mountains provide shelter from easterly winds. The driest area of Wales is in the north-east, in the lowlands bordering on Cheshire. Here there is a distinct rain shadow effect in the prevailing south-westerly maritime winds. To the south of this area, the country along the border with England is also subject to a rain shadow effect when the winds are westerly, but may experience considerable rainfall when depressions cross southern England.

The final distinct area is that of South Wales, from Pembrokeshire round to Glamorgan and southern Monmouthshire. This coastal area is generally mild, but tends to be wet and may be surprisingly cold in winter. The low-lying area around Cardiff, Newport and southern Monmouthshire is often affected in summer by fine weather over southern England, and may, at times, with southerly or south-westerly winds even derive some shelter from Exmoor on the other side of the Bristol Channel. In summer, warm conditions in southern England may also be transported west by easterly winds and affect the whole area. In winter, by contrast, the area may be cold as cold air drains out of the English Midlands along the Severn valley, bringing low stratus cloud or fog to the south-east of Wales.

7 Ireland

The climate of Ireland is dominated, as might be expected, by the Atlantic or, more specifically, by the warm waters of the North Atlantic Drift off its western coast. The whole island is subject to the maritime influence and is certainly far warmer than might otherwise be expected from its latitude. This is borne out by the extent, the quantity, and the quality of its grasslands. The length of time when the mean temperature exceeds 6°C, which is accepted as the limit for the growth of grass is exceptionally long. It is not for nothing that Ireland has earned the nickname of the 'Emerald Isle'. The climate is exceptionally equitable and there is very little variation in temperatures throughout the year. However, its location is firmly beneath one of the major storm tracks that are followed by depressions arriving from the Atlantic. By bearing the brunt

of any severe storms, not only does the island act as an 'early warning system' for the remainder of the British Isles, but also tends to temper their effects and reduce their severity for regions farther to the east.

Taken as a whole, the higher ground is located around the coasts of Ireland, with lower ground in the centre of the island. There is therefore a tendency for rainfall to be higher around the coasts than in the centre. One consequence of the warm sea is that in winter, in particular, the temperature difference between the cool land and the warm sea helps to strengthen developing depressions, so that frontal systems – and the rain that they bear – tend to be stronger in winter than in summer. Rainfall on hills close to the western coasts is greater at that time of the year, when there are also more showers, which add to the overall total amount of rain. This is particularly the case when unstable polar maritime air arrives in the wake of a depression. As it passes over the main flow of the North Atlantic Drift, it becomes strongly heated from below – sometimes by as much as 9 degrees Celsius – which thus increases its instability and the strength of the showers that are generated. All of which increases the rainfall.

As far as temperatures are concerned, there is very little variation across the island, although there tends to be a greater range in the north-east as compared with those prevailing over the south-west of Ireland. Distance from the coast also plays a part. The area with the greatest range of temperature is in southern Ulster, which experiences colder temperatures in late autumn and winter than areas along the southern and western coasts. From late spring until early autumn, maximum temperatures are higher in the north-east than in southern and western Ireland. In high summer, latitude does play a part, with temperatures being slightly higher in southern areas than along the north coast. One consequence of the equable maritime climate is that high temperatures, even in high summer, are rare, especially when compared with those that are experienced in southern England. On very rare occasions, in winter, the frigid air of the Siberian High may extend right across England and into Ireland and remains strong enough to overcome the influence of the warm air from the Atlantic. This was the case in 1962/1963 and in 2009/2010, but these winters were exceptional. Otherwise, frost is rare in coastal areas and occurs on just some 40 occasions in inland regions. As may be expected, snowfall is rare. Only in the extreme north-east does snow fall on an average of 30 days a year. In the far south-west, this figure decreases to about 5 days a year, and nowhere does snow lie for more than about a day or so.

The difference between the western coasts and the more sheltered inland areas is perhaps most obvious when wind strengths are compared. The strongest winds are observed on the northern coast of Ulster, which tends to be close to the area of the Atlantic over which depressions may undergo explosive deepening, with a consequent increase in wind speed. On rare occasions, such as the 'Night of the Big Wind', which began on the afternoon of 6 January 1839, highly destructive winds may extend right across the island, even into the far south.

8 Scotland

As with Wales, Scotland consists of several areas with diverse climates, which are a result of its mountainous as well as its maritime nature. Again as in the case with Wales, there are five distinct areas. Because they are at a distance from the mainland, the three island areas of the Hebrides (or Western Isles), Orkney and Shetland form one climatic area. In the west of the mainland, there is a distinction between the very mountainous Western Highlands region, running from Sutherland in the north, right the way down the west coast. The climate here is extremely wet because of the mountainous nature. To the east of this region, particularly in Caithness, Moray, and most of Aberdeenshire, although still a highland region (the Eastern Highlands), the area is shielded from the prevailing westerly winds by the Cairngorm Mountains and warmed by the Föhn effect. It is, however, fully exposed to frigid northerly winds, especially in the north, in Caithness, and particularly so in winter. On the east coast farther to the south, the climate is essentially the same from the Moray Firth, down the coastal strip of Aberdeenshire, Angus and Fife, across the Firth of Forth and as far as the eastern end of the Borders. Farther inland, the Central Lowlands and the western area of the Borders in Dumfries and Galloway and Ayrshire form yet another climatic region.

The highest temperatures are recorded in all areas in July (sometimes July and August in the outer islands), and there is a difference of some 2–3 degrees Celsius between the average temperature recorded in the Borders in the south and that found in the extreme north of the Eastern Highlands region (in Caithness). It is striking that the highest December and January temperatures in the whole of Britain have been recorded in northern Scotland. In both cases the temperature was 18.3°C, and both occurred as a result of the Föhn effect in the lee of high mountains. On 2 December 1948, Ashnaschellach in the mountainous Western Highlands region experienced this temperature and on 26 January 2003 it was Aboyne in the Eastern Highlands region in Aberdeenshire. High temperatures tend to occur in the Western Highlands with southerly or south-easterly winds, whereas the highest in the Eastern Highlands occur with westerly winds. The lowest temperatures occur in December or January in all regions with the exception of the Hebrides, Orkney and Shetland, where the lowest temperatures are recorded in February. It is believed that lower temperatures than those recorded at Braemar on 11 February 1893 and again on 10 January 1982 and at Altnaharra on 30 December 1990 may have occurred at other locations, where there are no recording stations.

Scotland is known for its wet climate. There is, however, quite a striking difference between rainfall in the west (about 1250 mm per year in the outer islands and even more in the Western Highlands region), and rainfall in the eastern coastal region (that running from the Moray Firth down to the Borders). Here, yearly totals of just 650–750 mm are typical. This key feature of the Scottish climate is, of course, related to the mountainous nature of the land on the west, which receives most of the precipitation and shields the rest of the country. Precipitation also falls as snow and, once again, there is a distinct difference between the west and east of the country.

Snowfall is also strongly dependent on altitude and here again the Western Highlands region receives greater amounts of snow than elsewhere. In the Western Isles and particularly in western Ayrshire, days on which snow is lying are very few.

When it comes to sunshine, the eastern coastal strip is favoured, as well as the area of the western edge of the Borders in the south. The sunniest areas in Scotland are northern Fife, the northern shore of the Firth of Tay and the Mull of Galloway. Here, sunshine totals may even match those found in southern England.

Thermopause
The transitional layer between the underlying mesosphere and the overlying exosphere. It is poorly defined and its altitude lies between 200 and 700 km, depending on solar activity.

Clouds

Altocumulus clouds.

Altocumulus

Altocumulus clouds (Ac) are medium-level clouds, with bases at 2–6 km, that, like all other varieties of cumulus, occur as individual, rounded masses. Although they may appear in small, isolated patches, they are normally part of extensive cloud sheets or layers and frequently form when gentle convection occurs within a layer of thin altostratus (page 234) breaking it up into separate heaps or rolls of cloud.

Altocumulus clouds may also take on the appearance of flat 'pancakes', but whatever the shape of the individual cloudlets, they always show some darker shading, unlike cirrocumulus. Blue sky is often visible between the separate masses of cloud – at least in the nearer parts of the layer.

Layers of altocumulus move as a whole, carried by the general wind at their height, but wind shear often causes the cloudlets to become arranged in long rolls or billows, which usually lie across the wind direction. High altocumulus or cirrocumulus of this type give rise to beautiful clouds that are commonly known as 'mackerel skies'.

Altostratus

Altostratus (As) is a dull, medium-level white or bluish-grey cloud in a relatively featureless layer, which may cover all or part of the sky. When illuminated by the rising or setting Sun, gentle undulations on the base may be seen, but these should not be confused with the regular ripples that often occur in altocumulus (page 233).

As with stratus, altocumulus may be created by gentle uplift. This frequently occurs at a warm front, where initial cirrostratus thickens and becomes altostratus, and the latter may lower and become rain-bearing nimbostratus (page 241). Patches or larger areas of altostratus may remain behind fronts, shower clouds or larger, organised storms. Conversely, altostratus may break up into altocumulus. Convection may then eat away at the cloud, until nothing is left.

Altostratus.

Cirrus.

Cirrus

Cirrus (Ci) is a wispy, thread-like cloud that normally occurs high in the atmosphere. Usually white, it may seem grey when seen against the light if it is thick enough.

Cirrus consists of ice crystals that are falling from slightly denser heads where the crystals are forming. In most cases wind speeds are higher at upper levels, so the heads move rapidly across the sky, leaving long trails of ice crystals behind them. Occasionally, the crystals fall into a deep layer of air moving at a steady speed. This can produce long, vertical trails of cloud.

Cirrocumulus

Cirrocumulus (Cc) is a very high white or bluish-white cloud, consisting of numerous tiny tufts or ripples, occurring in patches or larger layers that may cover a large part of the sky. The individual cloud elements are less than 1° across. They are sometime accompanied by fallstreaks of falling ice crystals. Unlike altocumulus (page 233) the small cloud elements do not show any shading. They are outlined by darker regions where the clouds are very thin or completely missing.

Cirrocumulus.

The cloud layer is often broken up into a regular pattern of ripples and billows. Clouds of this sort are often called a 'mackerel sky', although the term is sometimes applied to fine, rippled altocumulus. In fact, the differences between cirrocumulus and altocumulus are really caused only because the latter are lower and thus closer to the observer.

Cirrostratus
Cirrostratus (Cs) is a thin sheet of ice-crystal cloud and is most commonly observed ahead of the warm front of an approaching depression, when it is often the second cloud type to be noticed, after individual cirrus streaks. On many occasions, however, cirrostratus occurs as such a thin veil that it goes unnoticed, at least initially, until one becomes aware that the sky has lost its deep blue colour and has taken on a slightly milky appearance. The Sun remains clearly visible (and blindingly bright) through the cloud, but as the cirrostratus thickens, a slight drop in temperature may become apparent.

A solar halo in thin, almost invisible, cirrostratus cloud.

Once you realise that cirrostratus is present, it pays to check the sky frequently, because as it thickens it may display striking halo phenomena. This stage often passes fairly quickly as the cloud continues to thicken and lower towards the surface, eventually turning into thin altostratus (page 234).

Cirrostratus often has a fibrous appearance, especially if it arises from the gradual increase and thickening of individual cirrus streaks. Because the cloud is so thin, and contrast is low, the fibrous nature is easier to see when the Sun is hidden by lower, denser clouds or behind some other object.

Cumulus.

Cumulus

Cumulus clouds (Cu) are easy to recognise. They are the fluffy clouds that float across the sky on a fine day, and are often known as 'fair-weather clouds'. The individual heaps of cloud are generally well separated from one another – at least in their early stages. They have rounded tops and flat, darker bases. It is normally possible to see that these bases are all at one level. Together with stratus (page 241) and stratocumulus (page 242) they form closer to the ground than other cloud types.

The colour of cumulus clouds, like that of most other clouds, depends on how they are relative to the Sun and the observer. When illuminated by full sunlight, they are white – often blindingly white – but when seen against the Sun, unless they are very thin, they are various shades of grey.

Cumulonimbus

Cumulonimbus (Cb) is the largest and most energetic of the cumulus family. It appears as a vast mass of heavy, dense-looking cloud that normally reaches high into the sky. Its upper portion is usually brilliantly white in the sunshine, whereas its lower portions are very dark grey. Unlike the flat base of a cumulus, the bottom of a cumulonimbus is often ragged and it may even reach down to just above the ground. Shafts of precipitation are frequently clearly visible.

Cumulonimbus clouds consist of enormous numbers of individual convection cells, all growing rapidly up into the sky. Although cumulonimbus develop from tall cumulus, a critical difference is that at least part of their upper portions has changed from a hard 'cauliflower' appearance to a softer, more fibrous look. This is a sign that freezing has begun in the upper levels of the cloud.

Cumulonimbus.

Nimbostratus.

Stratus.

Nimbostratus

Nimbostratus (Nb) is a heavy, dark grey cloud with a very ragged
base. It is the main rain-bearing cloud in many frontal systems.
Shafts of precipitation (rain, sleet or snow) are visible below the
cloud, which is often accompanied by tattered shreds of cloud
that hang just below the base.

Just as cirrostratus often thickens and grades imperceptibly
into altostratus, so the latter may thicken into nimbostratus.
Once rain actually begins, or shafts of precipitation are seen to
reach the ground, it is safe to call the cloud nimbostratus.

Stratus

Stratus (St) is grey, water-droplet cloud that usually has a fairly
ragged base and top. It is always low and frequently shrouds
the tops of buildings. Indeed it is identical to fog, which may
be regarded as stratus at ground level. Although the cloud
may be thin enough for the outline of the Sun to be seen
clearly through it, in general it does not give rise to any optical
phenomena. It forms under stable conditions and is one of the
cloud types associated with 'anticyclonic gloom'.

Stratus may form either by gentle uplift (like the other
stratiform clouds) or when nearly saturated air is carried by a
gentle wind across a cold surface, which may be either land or
sea. Normally low wind speeds favour its occurrence, because
mixing is confined to a shallow layer near the ground. When
there is a large temperature difference between the air and the
surface, however, stratus may still occur, even with very strong
winds. Stratus also commonly forms when a moist air-stream
brings a thaw to a snow-covered surface.

There is very little precipitation from stratus, but it may
produce a little drizzle or, when conditions are cold enough,
even a few snow or ice grains. Ragged patches of stratus,
called 'scud' by sailors, often form beneath rain clouds such as
nimbostratus or cumulonimbus, especially where the humid air
beneath the rain cloud is forced to rise slightly, such as when
passing over low hills.

Stratocumulus

Stratocumulus (Sc) is a low, grey or whitish sheet of cloud, but unlike stratus it has a definite structure. There are distinct, separate masses of cloud that may be in the form of individual clumps, broader 'pancakes', or rolls. Sometimes these may be defined by thinner (and thus whitish) regions of cloud, but frequently blue sky is clearly visible between the masses of cloud.

Stratocumulus indicates stable conditions, and only slow changes to the current weather. It generally arises in one of two ways: either from the spreading out of cumulus clouds that reach an inversion, or through the break-up of a layer of stratus cloud. In the first case, the tops of the cumulus flatten and spread out sideways when they reach the inversion, producing clouds that are fairly even in thickness, with flat tops and bases. Initially, perhaps early in the day, there may be large areas of clear air, but the individual elements gradually merge to cover a larger area, or even completely blanket the sky.

In the second case, shallow convection (whose onset is often difficult to predict) begins within a sheet of stratus, causing the layer to break up. The regions of thinner cloud or clear air indicate where the air is descending, and the thicker, darker centres where it is rising.

Stratocumulus.

Cloud heights

The height of clouds is usually given in feet (often with approximate metric equivalents). This may seem odd, when all other details of clouds, and meteorology in general, uses metric (SI) units. It is, however, a hangover from the way in which aircraft heights are specified. When aviation became general between the two World Wars, most commercial flying took place in the United Kingdom and America, so heights of aircraft and airfields were given in feet. It was obviously essential for cloud heights to be the same. The practice has continued: the heights of airfields, aircraft and clouds are still given in feet. The World Meteorological Organization recognises three ranges of cloud heights: low, middle and high. Clouds are specified by the height of their bases, not by that of their tops. The three divisions are:

Low clouds (bases 6500 feet or lower, approx. 2 km and below): cumulus (page 238), stratocumulus (page 242), stratus (page 241).

Middle clouds (bases between 6500 and 20,000 feet, approx. 2 to 6 km): altocumulus (page 233), altostratus (page 234), nimbostratus (page 241).

High clouds (bases over 20,000 feet, above 6 km): cirrus (page 235), cirrocumulus (page 235), cirrostratus (page 236).

One cloud type, cumulonimbus (page 239) commonly stretches through all three height ranges. Nimbostratus, although nominally a middle-level cloud, is frequently very deep and although it has a low base, may extend to much higher altitudes.

Tornadoes

The most intense and damaging tornadoes are those produced beneath supercells. A supercell consists of a rotating system of air, a mesocyclone, with low pressure in the centre. In the northern hemisphere this will rotate cyclonically, that is, anticlockwise. The rotation of the clouds in a mesocyclone is often clearly visible in their structure. In general, a visible condensation funnel descends from the cloud, to give a rotating wall cloud. If this funnel reaches the surface it becomes a true tornado, capable of lifting a rotating column of debris and dust from the ground.

The descent of this funnel, and thus the formation of tornadoes, is poorly understood. One suggestion is that the initial stage of a tornado consists of a horizontal rotating tube of air beneath a mesocyclone. The centre of this tube is raised by the upcurrents beneath the mesocyclone to produce an 'inverted U-shaped' rotating tube of air. One limb is enhanced by the strong upcurrents of the overlying cloud, while the other limb is supressed, resulting in the final tornado form.

Generally, the speed of the wind in a tornado is about 180 kph, although the strongest are known to reach over 300 kph and are capable of lifting several tonnes at a time. The general diameter is about 80 metres although again this can be greatly exceeded; they are known to reach a diameter of 3 km. The lifetime is generally short, although a true tornado may travel on the ground for several kilometres. The largest are known to stay on the surface for some 100 km, causing immense destruction along their track. The longest known tornado track in Britain occurred on 21 May 1950, when the tornado is believed to have tracked from Little London in Buckinghamshire to Coveny in Cambridgeshire, a distance of just over 107 km. It then lifted into a funnel cloud and travelled another 52.6 km to Shipham in Norfolk, disappearing over the North Sea.

Because of this capacity of widespread damage, predicting tornadoes is of vital importance. Modern advances in doppler radar have improved forecasters' ability to detect mesocyclones

One of the earliest photographs of a tornado, taken in Kansas in 1884.

as they are forming, but it remains difficult to predict if, when and where these storms will develop into tornadoes. Warning times in 'tornado alley' in the south-eastern USA have got shorter, but are still rarely more than 15 minutes. More reassuringly, they have got more accurate, with fewer false alarms and 87 per cent of deadly tornadoes occurring with advance warning.

A far more recent tornado picks up debris in the American Midwest.

Although people often speak of waterspouts as 'tornadoes over water', in fact waterspouts (and the related landspouts, although that term is rarely used) form by a completely different mechanism. In these, the up- and down-draughts from a vigorous cumulonimbus or even cumulus congestus cloud are sufficiently strong to reach the surface, at which point the up-draughts lift material into the air.

The Beaufort Scale

Wind strength is commonly given on the Beaufort scale. This was originally defined by Francis Beaufort (later Admiral Beaufort) for use at sea, but was subsequently modified for use on land. Meteorologists generally specify the speed of the wind in metres per second (m s⁻¹). For wind speeds at sea, details are usually given in knots. The equivalents in kph are shown for speeds over land.

The Beaufort scale (for use at sea)

Force	Description	Sea state	Speed	
			Knots	m s⁻¹
0	calm	like a mirror	<1	0.0–0.2
1	light air	ripples, no foam	1–3	0.3–1.5
2	light breeze	small wavelets, smooth crests	4–6	1.6–3.3
3	gentle breeze	large wavelets, some crests break, a few white horses	7–10	3.4–5.4
4	moderate breeze	small waves, frequent white horses	11–16	5.5–7.9
5	fresh breeze	moderate, fairly long waves, many white horses, some spray	17–21	8.0–10.7
6	strong breeze	some large waves, extensive white foaming crests, some spray	22–27	10.8–13.8

The Beaufort scale (for use at sea) – *continued*

Force	Description	Sea state	Speed	
			Knots	m s⁻¹
7	near gale	sea heaping up, streaks of foam blowing in the wind	28–33	13.9–17.1
8	gale	fairly long and high waves, crests breaking into spindrift, foam in prominent streaks	34–40	17.2–20.7
9	strong gale	high waves, dense foam in wind, wave-crests topple and roll over, spray interferes with visibility	41–47	20.8–24.4
10	storm	very high waves with overhanging crests, dense blowing foam, sea appears white, heavy tumbling sea, poor visibility	48–55	24.5–28.4
11	violent storm	exceptionally high waves may hide small ships, sea covered in long, white patches of foam, waves blown into froth, poor visibility	56–63	28.5–32.6
12	hurricane	air filled with foam and spray, visibility extremely bad	64	32.7

The Beaufort scale (adapted for use on land)

Force	Description	Events on land	Speed km h⁻¹	m s⁻¹
0	calm	smoke rises vertically	<1	0.0–0.21
1	light air	direction of wind shown by smoke but not by wind vane	1–5	0.3–1.5
2	light breeze	wind felt on face, leaves rustle, wind vane turns to wind	6–11	1.6–3.3
3	gentle breeze	leaves and small twigs in motion, wind spreads small flags	12–19	3.4–5.4
4	moderate breeze	wind raises dust and loose paper, small branches move	20–29	5.5–7.9
5	fresh breeze	small leafy trees start to sway, wavelets with crests on inland waters	30–39	8.0–10.7
6	strong breeze	large branches in motion, whistling in telephone wires, difficult to use umbrellas	40–50	10.8–13.8
7	near gale	whole trees in motion, difficult to walk against wind	51–61	13.9–17.1

The Beaufort scale (adapted for use on land) – *continued*

Force	Description	Events on land	Speed km h⁻¹	m s⁻¹
8	gale	twigs break from trees, difficult to walk	62–74	17.2–20.7
9	strong gale	slight structural damage to buildings; chimney pots, tiles and aerials removed	75–87	20.8–24.4
10	storm	trees uprooted, considerable damage to buildings	88–101	24.5–28.4
11	violent storm	widespread damage to all types of building	102–117	28.5–32.6
12	hurricane	widespread destruction, only specially constructed buildings survive	≥118	32.7≥

The TORRO Tornado Scale

The TORRO tornado intensity scale is based on an extension to the Beaufort scale of wind speeds. The winds speeds are actually calculated mathematically from the accepted Beaufort wind speeds. (Although the Beaufort scale was first proposed in 1805, it was expressed in terms of wind speed in 1921.) T0 corresponds to Beaufort Force 8, and T11 would correspond to Beaufort Force 30 (if such a force existed).

The TORRO scale is thus solely based on wind speeds, unlike the Fujita scale and the later, modified version, the Enhanced Fujita scale, which are based on an assessment of damage. In practice, wind-speed measurements are rarely available for tornadoes, and so, in effect, both scales are, perforce, based on an assessment of the intensity of damage.

Scale	Wind speed (estimated)			Potential damage
	mph	km h⁻¹	m s⁻¹	
F0	0–38	0–60	0–16	**No damage.** *(Funnel cloud aloft, not a tornado)* No damage to structures, unless on tops of tallest towers, or to radiosondes, balloons and aircraft. No damage in the country, except possibly agitation to highest tree-tops and effect on birds and smoke. A whistling or rushing sound aloft may be noticed.
T0	39–54	61–86	17–24	**Light damage.** Loose light litter raised from ground-level in spirals. Tents, marquees seriously disturbed; most exposed tiles, slates on roofs dislodged. Twigs snapped; trail visible through crops.
T1	55–72	87–115	25–32	**Mild damage.** Deckchairs, small plants, heavy litter becomes airborne; minor damage to sheds. More serious dislodging of tiles, slates, chimney pots. Wooden fences flattened. Slight damage to hedges and trees.

The TORRO Tornado Scale – *continued*

Scale	Wind speed (estimated)			Potential damage
	mph	km h⁻¹	m s⁻¹	
T2	73–92	116–147	33–41	**Moderate damage.** Heavy mobile homes displaced, light caravans blown over, garden sheds destroyed, garage roofs torn away. Much damage to tiled roofs and chimney stacks. General damage to trees, some big branches twisted or snapped off, small trees uprooted.
T3	93–114	148–184	42–51	**Strong damage.** Mobile homes overturned / badly damaged; light caravans destroyed; garages and weak outbuildings destroyed; house roof timbers considerably exposed. Some larger trees snapped or uprooted.
T4	115–136	185–220	52–61	**Severe damage.** Motor cars levitated. Mobile homes airborne / destroyed; sheds airborne for considerable distances; entire roofs removed from some houses; roof timbers of stronger brick or stone houses completely exposed; gable ends torn away. Numerous trees uprooted or snapped.
T5	137–160	221–259	62–72	**Intense damage.** Heavy motor vehicles levitated; more serious building damage than for T4, yet house walls usually remaining; the oldest, weakest buildings may collapse completely.

The TORRO Tornado Scale – *continued*

Scale	Wind speed (estimated)			Potential damage
	mph	km h⁻¹	m s⁻¹	

Scale	mph	km h⁻¹	m s⁻¹	Potential damage
T6	161–186	260–299	73–83	**Moderately-devastating damage.** Strongly built houses lose entire roofs and perhaps also a wall; windows broken on skyscrapers, more of the less-strong buildings collapse.
T7	187–212	300–342	84–95	**Strongly-devastating damage.** Wooden-frame houses wholly demolished; some walls of stone or brick houses beaten down or collapse; skyscrapers twisted; steel-framed warehouse-type constructions may buckle slightly. Locomotives thrown over. Noticeable debarking of trees by flying debris.
T8	213–240	343–385	96–107	**Severely-devastating damage.** Motor cars hurled great distances. Wooden-framed houses and their contents dispersed over long distances; stone or brick houses irreparably damaged; skyscrapers badly twisted and may show a visible lean to one side; shallowly anchored high rises may be toppled; other steel-framed buildings buckled.
T9	241–269	386–432	108–120	**Intensely-devastating damage.** Many steel-framed buildings badly damaged; skyscrapers toppled; locomotives or trains hurled some distances. Complete debarking of any standing tree-trunks.

The TORRO Tornado Scale – *continued*

Scale	Wind speed (estimated)			Potential damage
	mph	km h⁻¹	m s⁻¹	
T10	270–299	433–482	121–134	**Super damage.** Entire frame houses and similar buildings lifted bodily or completely from foundations and carried a large distance to disintegrate. Steel-reinforced concrete buildings may be severely damaged or almost obliterated.
T11	>300	>483	>135	**Phenomenal damage.** Strong framed, well-built houses levelled off foundations and swept away. Steel-reinforced concrete structures are completely destroyed. Tall buildings collapse. Some cars, trucks and train carriages may be thrown approximately 1 mile (1.6 kilometres).

TORRO Hailstorm Intensity Scale

The Tornado and Storm Research Organisation (TORRO) has not only developed a scale for rating tornadoes (see pages 252–255) but also one to judge the severity of hailstorm incidents. This scale is given in the following table, but it must be borne in mind that the severity of any hailstorm will depend (among other factors) upon the size of individual hailstones, their numbers and also the speed at which the storm itself travels across country.

Scale	Intensity	Hail size (mm)	Size comparison	Damage
H0	Hard hail	5–9	Pea	None
H1	Potentially damaging	10–15	Mothball	Slight general damage to plants, crops
H2	Significant	16–20	Marble, grape	Significant damage to fruit, crops, vegetation
H3	Severe	21–30	Walnut	Severe damage to fruit and crops Damage to glass and plastic structures Paint and wood scored
H4	Severe	31–40	Pigeon's egg > squash ball	Widespread damage to glass Damage to vehicle bodywork
H5	Destructive	41–50	Golf ball > pullet's egg	Wholesale destruction of glass Damage to tiled roofs Significant risk of injuries

TORRO Hailstorm Intensity Scale – *continued*

Scale	Intensity	Hail size (mm)	Size comparison	Damage
H6	Destructive	51–60	Hen's egg	Bodywork of grounded aircraft dented Brick walls pitted
H7	Destructive	61–75	Tennis ball > cricket ball	Severe roof damage Risk of serious injuries
H8	Destructive	76–90	Large orange > soft ball	Severe damage to aircraft bodywork
H9	Super hailstorms	91–100	Grapefruit	Extensive structural damage
H10	Super hailstorms	>100	Melon	Extensive structural damage Risk of severe or fatal injuries to persons caught in the open

Twilight Diagrams

Sunrise, sunset, twilight

For each individual month, we give details of sunrise and sunset times for the four capital cities of the various countries that make up the United Kingdom.

During the summer, especially at high latitudes, twilight may persist throughout the night and make it difficult to see the faintest stars. Beyond the Arctic and Antarctic Circles, of course, the Sun does not set for 24 hours at least once during the summer (and rise for 24 hours at least once during the winter). Even when the Sun does dip below the horizon at high latitudes, bright twilight persists throughout the night, so observing the fainter stars is impossible. Even in Britain this applies to northern Scotland, which is why we include a diagram for Lerwick in the Shetland Islands.

As mentioned earlier (page 9) there are three recognised stages of twilight: civil twilight, nautical twilight and astronomical twilight. Full darkness occurs only when the Sun is more than 18° below the horizon. During nautical twilight, only the very brightest stars are visible. During astronomical twilight, the faintest stars visible to the naked eye may be seen directly overhead, but are lost at lower altitudes. They become visible only once it is fully dark. The diagrams show the duration of twilight at the various locations. Of the locations shown, during the summer months there is astronomical twilight for a short time at Belfast, and this lasts longer during the summer at all of the other locations. To illustrate the way in which twilight occurs in the far south of Britain, we include a diagram showing twilight duration at St Mary's in the Scilly Isles. (A similar situation applies to the Channel Islands, which are also in the far south.) Once again, full darkness never occurs.

The diagrams show the times of New and Full Moon (black and white symbols, respectively). As may be seen, at most locations during the year roughly half of New and Full Moon phases may come during daylight. For this reason, the exact phase may be invisible in Britain, but be clearly seen elsewhere in the world. The exact times of the events are given in the diagrams for each individual month.

Lerwick, Shetland Islands – Latitude 60.2°N – Longitude 1.1°W

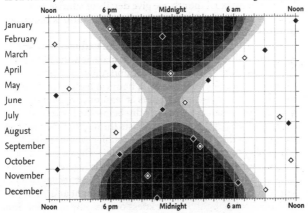

Edinburgh, UK – Latitude 55.9°N – Longitude 3.2°W

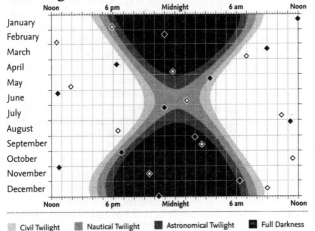

Civil Twilight · Nautical Twilight · Astronomical Twilight · Full Darkness

◇ Time of Full Moon · ◆ Time of New Moon

Belfast, UK – Latitude 54.6°N – Longitude 5.8°W

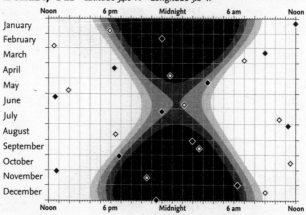

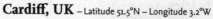

Cardiff, UK – Latitude 51.5°N – Longitude 3.2°W

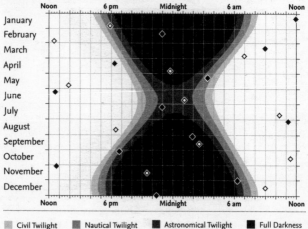

Civil Twilight   Nautical Twilight   Astronomical Twilight   Full Darkness

◇ Time of Full Moon   ◆ Time of New Moon

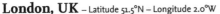

London, UK – Latitude 51.5°N – Longitude 2.0°W

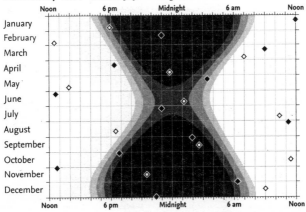

St Mary's, Scilly Isles – Latitude 49.9°N – Longitude 6.4°W

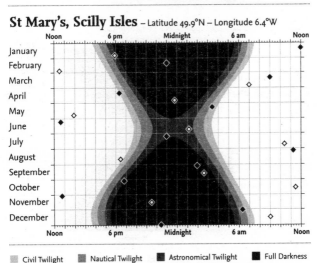

Civil Twilight   Nautical Twilight   Astronomical Twilight   Full Darkness
◇ Time of Full Moon   ◆ Time of New Moon

Further Reading and Internet Links

Books

Chaboud, René, *How Weather Works* (Thames & Hudson, 1996)

Dunlop, Storm, *Clouds* (Haynes, 2018)

Dunlop, Storm, *Collins Gem Weather* (HarperCollins, 1999)

Dunlop, Storm, *Collins Nature Guide Weather* (HarperCollins, 2004)

Dunlop, Storm, *Come Rain or Shine* (Summersdale, 2011)

Dunlop, Storm, *Dictionary of Weather* (2nd edition, Oxford University Press, 2008)

Dunlop, Storm, *Guide to Weather Forecasting* (rev. printing, Philip's, 2013)

Dunlop, Storm, *How to Identify Weather* (HarperCollins, 2002)

Dunlop, Storm, *How to Read the Weather* (Pavilion, 2018)

Dunlop, Storm, *Weather* (Cassell Illustrated, 2006/2007)

Eden, Philip, *Weatherwise* (Macmillan, 1995)

File, Dick, *Weather Facts* (Oxford University Press, 1996)

Hamblyn, Richard & Meteorological Office, *The Cloud Book: How to Understand the Skies* (David & Charles, 2009)

Hamblyn, Richard & Meteorological Office, *Extraordinary Clouds* (David & Charles, 2009)

Kington, John, *Climate and Weather* (HarperCollins, 2010)

Ludlum, David, *Collins Wildlife Trust Guide Weather* (HarperCollins, 2001)

Meteorological Office, *Cloud Types for Observers* (HMSO, 1982)

Met Office, Factsheets 1–19 (pdfs downloadable from: http://www.metoffice.gov.uk/learning/library/publications/factsheets)

Watts, Alan, *Instant Weather Forecasting* (Adlard Coles Nautical, 2000)

Watts, Alan, *Instant Wind Forecasting* (Adlard Coles Nautical, 2001)

Watts, Alan, *The Weather Handbook* (3rd edn, Adlard Coles Nautical, 2014)

Whitaker, Richard, ed., *Weather: The Ultimate Guide to the Elements* (HarperCollins, 1996)

Williams, Jack, *The AMS Weather Book: The Ultimate Guide to America's Weather* (Univ. Chicago Press, 2009)

Woodward, A., & Penn, R., *The Wrong Kind of Snow* (Hodder & Stoughton, 2007)

Internet links – Current weather

AccuWeather: *http://www.accuweather.com/*
     UK: *http://www.accuweather.com/ukie/index.asp?*

Australian Weather News:
    http://www.australianweathernews.com/

    UK station plots:
      http://www.australianweathernews.com/sitepages/
      charts/611_United_Kingdom.shtml

BBC Weather: *http://www.bbc.co.uk/weather*

CNN Weather: *http://www.cnn.com/WEATHER/index.html*

Intellicast: *http://intellicast.com/*

ITV Weather: *http://www.itv-weather.co.uk/*

Unisys Weather: *http://weather.unisys.com/*

UK Met Office: *http://www.metoffice.gov.uk*

    Forecasts:
      *http://www.metoffice.gov.uk/weather/uk/uk_forecast_*
      weather.html

    Hourly Weather Data:
      http://www.metoffice.gov.uk/education/teachers/
      latest-weather-data-uk

    Latest station plot:
      http://www.metoffice.gov.uk/data/education/chart_latest.gif

    Surface pressure charts:
      http://www.metoffice.gov.uk/public/weather/surface-pressure/

    Explanation of symbols on pressure charts:
      http://www.metoffice.gov.uk/guide/weather/
      symbols#pressure-symbols

    Synoptic & climate stations (interactive map):
      http://www.metoffice.gov.uk/public/weather/climate-network/
      #?tab=climateNetwork

    Weather on the Web:
      http://wow.metoffice.gov.uk/

The Weather Channel:
    http://www.weather.com/twc/homepage.twc

Weather Underground:
    http://www.wunderground.com

Wetterzentrale: *http://www.wetterzentrale.de/pics/Rgbsyn.gif*

Wetter3 (German site with global information):
    http://www.wetter3.de

    UK Met Office chart archive:
      http://www.wetter3.de/Archiv/archiv_ukmet.html

General information

Atmospheric Optics:
http://www.atoptics.co.uk/

Hurricane Zone Net:
http://www.hurricanezone.net/

National Climate Data Centre:
http://www.ncdc.noaa.gov/

Extremes:
http://www.ncdc.noaa.gov/oa/climate/severeweather/extremes.html

National Hurricane Center:
http://www.nhc.noaa.gov/

Reading University (Roger Brugge):
http://www.met.reading.ac.uk/~brugge/index.html

UK Weather Information:
http://www.weather.org.uk/

Unisys Hurricane Data:
http://weather.unisys.com/hurricane/atlantic/index.html

WorldClimate:
http://www.worldclimate.com/

Meteorological Offices, Agencies and Organisations

Environment Canada:
http://www.msc-smc.ec.gc.ca/

European Centre for Medium-Range Weather Forecasting (ECMWF):
http://www.ecmwf.int

European Meteorological Satellite Organisation:
http://www.eumetsat.int/website/home/index.html

Intergovernmental Panel on Climate Change:
http://www.ipcc.ch

National Oceanic and Atmospheric Administration (NOAA):
http://www.noaa.gov/

National Weather Service (NWS):
http://www.nws.noaa.gov/

UK Meteorological Office:
http://www.metoffice.gov.uk

World Meteorological Organisation:
http://www.wmo.int/pages/index_en.html

Satellite images

Eumetsat:
> *http://www.eumetsat.de/*
>> Image library:
>> *http://www.eumetsat.int/website/home/Images/ImageLibrary/index.html*

Group for Earth Observation (GEO):
> *http://www.geo-web.org.uk/*

Societies

American Meteorological Society:
> *http://www.ametsoc.org/AMS*

Australian Meteorological and Oceanographic Society:
> *http://www.amos.org.au*

Canadian Meteorological and Oceanographic Society:
> *http://www.cmos.ca/*

Climatological Observers Link (COL):
> *https://colweather.ssl-01.com/*

European Meteorological Society:
> *http://www.emetsoc.org/*

Irish Meteorological Society:
> *http://www.irishmetsociety.org*

National Weather Association, USA:
> *http://www.nwas.org/*

New Zealand Meteorological Society:
> *http://www.metsoc.org.nz/*

Royal Meteorological Society:
> *http://www.rmets.org*

TORRO: Hurricanes and Storm Research Organisation:
> *http://torro.org.uk*

Acknowledgements

28	The loss of the Pennsylvania	Royal Museums Greenwich/ Wikimedia Commons
29	Newspaper report of the Night of the Big Wind	*Northern Whig*
45	Dawlish sea wall	BBC
60	Higher Shelf Stones	Shutterstock
61	Snake Pass	Shutterstock
76	Robert FitzRoy's weather forecast	*The Times*
77	Francis Galton's weather map	*The Times*
78	Small tortoiseshell butterfly	Shutterstock
94	1955 Fan Blow	BBC
95	The Dust Bowl	Wikimedia Commons
98	*Flaming June*	World History Archive/ Alamy Stock Photo
111	Map of D-Day landings	Shutterstock
113	Noctilucent clouds	Alan Tough
129	HMS Association	Royal Museums Greenwich/Wikimedia Commons
133	Robert Fitzroy	Liam White/Alamy Stock Photo
145	Insects by Jan van Kessel	Ashmolean Museum/ Wikimedia Commons
160	London on fire with St Pauls on the horizon	Shutterstock
161	London Bridge on fire	Hulton Archive/Getty images
177	The crater of Vesuvius	Wikimedia Commons (CC BY-SA 3.0)
192	The successive Eddystone lighthouses	Trinity House
193	Winstanley's second lighthouse	State Library of New South Wales/Wikimedia Commons
194	The modern lighthouse	Wikimedia Commons
208	Mr Pickwick sliding on the ice	*Pickwick Papers*, February 1837 via Victorian Web
233–243	Cloud types	Storm Dunlop
246	Early photograph of a tornado	Kansas State Historical Society
247	A more recent tornado	Shutterstock

Index